514

D0044614

Praise for *The Blue Man and Other Stories of the Skin*

"How does our skin define us? In *The Blue Man,* dermatologist-author Robert Norman not only provides an entertaining and informative overview of how our skin shields us from harm but also provides clues that disease may be running amok in our bodies. *The Blue Man* will appeal to all those who make the health and care of skin their business and to general readers seeking to learn more about the body's largest organ."

Rita Ciresi, author of *Pink Slip* and *Bring Back My Body to Me,* Professor of English and Director of Creative Writing, University of South Florida

"Lucid and appealing, *The Blue Man* is an introduction to a topic that almost all of us are interested in: our skin. Dr. Norman is a well-recognized expert in dermatology. For students interested in medical narrative, certain chapters would be required reading."

David J. Elpern, M.D., Editor, *The Online Journal of Community and Person-Centered Dermatology (OJCPCD)*

"Dr. Robert Norman's *The Blue Man* describes real-life dermatological detective stories that reveal the skin as a complex and mysterious creature and the practicing physician as both detective and epidemiologist. This book educates and entertains, illustrates and illuminates, and improves our understanding of the skin—our largest organ and the link between our bodies, the environment and other organisms. Read this book and enter the curious and fascinating world of skin."

Sharad P. Paul, M.D., author of *Skin: A Biography*

"Skin diseases may create great physical, mental, and emotional suffering. Dr. Robert Norman writes about these afflictions with insight and compassion."

John E. Wolf, Jr., M.D., M.A., Professor and Chairman, Department of Dermatology, Baylor College of Medicine

"Human skin, serving as the interface with the environment, is not only an important barrier, but it is also a major contributor to social and cultural interactions. Thus, in addition to considerations of disease, there is much to reflect on the skin's greater role in our lives. While the biomedically oriented physician may focus on making the right diagnosis and prescribing the right therapy, the humanistic physician sees the big picture, the stories of people's lives in which the skin plays a prominent role. Leading dermatologist Dr. Robert Norman—a man of science, letters, and sound—shares these stories, educating us on the basic science underlying this remarkable organ while demonstrating the lessons it has to share with us in our interpersonal lives."
Steven R. Feldman, M.D., Ph.D., Department of Dermatology, Wake Forest University School of Medicine

"Dr. Norman's stories about our skin range from its vital physiological functions to its behavior in our everyday lives. The book is a beautifully written account that celebrates the body's largest organ. Dr. Norman frequently uses the 'eye of a naturalist' to provide comparisons between the behavior of skin and what occurs around us in nature to emphasize his points. A delightful read!"
Irwin M. Braverman M.D., Professor Emeritus of Dermatology, Yale Medical School

"Dr. Norman is a born raconteur. He has taken his literary skills to new heights in gathering together stories about the skin and has capably used them to illustrate various dermatologic observations and findings."
Lawrence Charles Parish, M.D., M.D. (Hon.), Editor in Chief, *Clinics in Dermatology* and *SkinMED*

THE BLUE MAN
AND OTHER STORIES
OF THE SKIN

DR. ROBERT NORMAN

THE BLUE MAN

AND OTHER
STORIES
OF THE
SKIN

UNIVERSITY OF CALIFORNIA PRESS

Berkeley Los Angeles London

University of California Press, one of the most distin-
guished university presses in the United States, enriches
lives around the world by advancing scholarship in the
humanities, social sciences, and natural sciences. Its
activities are supported by the UC Press Foundation and
by philanthropic contributions from individuals and institutions. For more information, visit www.ucpress.edu.

University of California Press
Berkeley and Los Angeles, California

University of California Press, Ltd.
London, England

Library of Congress Cataloging-in-Publication Data

Norman, Robert A., 1955–.
    The blue man and other stories of the skin /
Robert A. Norman.
        p. ; cm.
    Includes bibliographical references.
    ISBN 978-0-520-27286-6 (cloth : alk. paper)
    1. Skin Physiological Phenomena.
2. Skin—pathology.   3. Skin Diseases.   I. Title
    RL96   2014
    616.5—dc23

                                        2013033771

Manufactured in the United States of America

23   22   21   20   19   18   17   16   15   14
10   9   8   7   6   5   4   3   2   1

The paper used in this publication meets the minimum
requirements of ANSI/NISO Z39.48–1992 (R 2002)
(Permanence of Paper).

For Howard

CONTENTS

PREFACE

To cure sometimes, to relieve often, to comfort always.

—Attributed to Edward Livingston Trudeau

Throughout my medical career, I have kept two sets of records: one of patients' clinical notes, the other of patients' stories. Along the way, I began to write my own stories. I have hammered out essays for *Discover* magazine, written and edited five textbooks of dermatology, and been a preceptor for a few hundred medical residents, medical students, and undergraduate students who ask endless questions and increase my own knowledge while feeding their own.

This book focuses on the dynamic qualities of the skin and the people who inhabit it. Included are stories about the life of the skin, tales of fascinating and mysterious patients, meditations on cultural issues including skin color, and thoughts about the future of the skin. I try to speak for the skin, since it cannot speak for itself, and show you how lucky we are to have such an amazing natural covering. Our skin is our partner, nourishing and protecting us, teaching us about the world, and serving as a record of our lives. We should take the opportunity to learn what the skin has to teach and how we can best serve each other.

Each day I try to heal the pain, both psychological and physical, that illness of the skin causes people. As a physician, I am often required to peel away possibilities to get to the source of my patient's medical problem and repair what has been disturbed or broken. Each problem—and each appointment—is preceded by a story. Recently a man came to see me complaining about a "growth on my arm giving me pain." In the exam room, my medical assistant gently pulled off the man's ragged, home-rigged dressing and revealed a profusely bleeding half-dollar-sized skin cancer. I let him know that I could surgically remove it right then, and he agreed. But beyond this nasty growth I could envision a whole history, including a lack of funds and transportation, that had kept him from seeking a remedy. A few days later an eight-year-old girl came to my office with severe atopic dermatitis and a horrible itch. She had been out of school for three weeks—and her mother had been unable to go to work during much of that time, due to lack of sleep and trying to care for her child. With each patient's story, the narrative proceeds to a resolution and hopefully some relief from the burden of a disease that elicits pain, itch, or other disturbance in quality of life for the patient and family.

On many days, my primary role is that of a nurturer. I often have to explain that many skin diseases are chronic in nature and only palliative care can be provided. Although there may be a perception that skin conditions only minimally affect patients, those who have protracted and severe conditions often endure serious psychosocial repercussions. All the activities of daily living—including work, sleep, hobbies, and social contact—may be harshly disturbed. Using the Dermatology Life Quality Index, the Beck Depression Inventory, and other quality-of-life measures, researchers have confirmed that those suffering from prolonged skin disease often endure significant impairments of work, school, and personal relationships.

In my stories and in my daily work, the patient is at the center, and the disease is a supporting actor. In the words of one of my heroes, William Carlos Williams, I get to "witness the words being born" while listening to my patients describe what bothers them and why they are seeking help. If I pay close attention and ask the pertinent questions, I can become a part of the birthing process; later, at my writing desk, I am able to record and perhaps better comprehend what has been said and done. In the exam room I'm confronted with themes from the marriage of medicine and literature, reflections on what is happening today in medicine and the world, and questions about how each person lives and interprets personal problems. Although these observations and conversations primarily take place during my exam of a patient, the history that precedes each problem has occurred over years or decades—my job is to take the kernel of a patient's story that I can glimpse in the exam room and make it come to life in my writing.

On the journey across these pages, I will ask many questions. What can we do to protect our skin and ourselves? How will the skin of the future be different? How does skin color affect social position and how can diseases such as albinism impose real punishment? How do people with skin diseases see themselves and their opportunities in life, and how do their lives change as a result of their conditions? How do people cope with having bodies or diseases that aren't "normal" in our culture? My overall purpose is to showcase the wonders of the skin, the agony of certain diseases and their attendant social and psychological toll, and provide some insight into how the skin and our perception of it influence our social, cultural, spiritual, and physical being—and how we can learn and grow from our new knowledge.

Why would someone choose dermatology as a career? In his introduction to *The Life of the Skin: What It Hides, What It Reveals, and How It Communicates,* by Arthur K. Balin and Loretta Pratt Balin, the famous dermatologist Albert Kligman wrote, "When I began, the field of dermatology was a swamp of snakes and nonsense, magic and folklore. The cynical view used to be that people went into dermatology so they didn't have to be awakened at night to go care for patients." He continues, "Modern dermatology is rooted in science and developing at an Olympic pace. These days the highest-ranking students apply for dermatology residencies. To become a dermatologist now, a student has to be highly educated and conversant in the fields of psychology, art, immunology, genetics, and biochemistry, to name but a few."

As the Balins carefully point out in their book, "Dermatologists do far more than treat the skin, for the skin mirrors the health of the entire body." By means of an eruption, itch, color change, or other sign, the skin can alert us to a problem in the deeper organs. For the visually oriented, the signs and patterns of skin disease signal the diagnosis. But dermatology is also tactile: I often guide the fingertips of medical students to help them touch the sandpaper-like feel of a precancerous skin lesion, the warmth of an area of inflammation, or the blanching of a vascular lesion.

Finally, as Kligman points out, dermatologists also have to know psychology. I see people at their most vulnerable, devastated as many are by their appearance. There is a gratification for both me and my patient from seeing the results of our work, either rapidly, when removing a skin cancer, or over time, when treating psoriasis or cystic acne.

I have divided the book into two parts. I start at the surface, with the enormously dynamic and lifesaving nature of the skin itself. Part 1, "Our Vital Skin," highlights the skin in its many amazing roles. I'll investigate how our skin defines us, what we need to do to protect it, what unseen creatures live on it, and how crucial structures within it have determined the course of history.

Part 2, "Living in Our Skins," goes deeper, into how we deal with skin diseases, the social and psychological issues of skin as social entertainment, and how the skin reveals and communicates our wishes and burdens. The focus here is on how skin color, disease, or irregularity can affect social position, self-esteem, and a person's opportunities in life. In this section, we'll see how our skin in all its glory and vagaries—from extra ears to imaginary parasites—manifests itself in our lives and culture.

Throughout, we'll see that skin has a direct cultural influence on every generation. It is a canopy for each individual and for all of humanity, stretching over us and encompassing the whole range of shape, color, and disturbance, from idiosyncratic, rare diseases to the broad experience of entire societies.

ACKNOWLEDGMENTS

Many thanks to Rachel Berchten and Christopher Lura, and all the editors at the University of California Press. Thanks also to Julie Christianson for her marketing acumen. And I appreciate very much the work of freelance editor Caroline Knapp. Special thanks to editor Naomi Schneider for her patience and for carefully offering great suggestions and helping me through this labor of love.

Great appreciation goes to my brother Howard, who has always helped me in trying to get this writing thing down with the highest quality I can manage and in bringing this book to fruition. Thanks to Dr. David Elpern and others who read earlier drafts of the book and provided useful insight and to those that offered endorsements. As always, I send much love to my family, friends, and patients for all I learn and share with them. Thanks to those that understand my drive for creation that sometimes cuts into the structure of our lives and time, but I hope my work in medicine and writing makes up a little for my absence.

PART ONE

OUR VITAL SKIN

Our skin mediates the most important transactions
of our lives. Skin is key to our biology, our sensory
experiences, our information gathering, and our relationships with others. Although the many roles it plays
are rarely appreciated, it is one of the most remarkable
and highly versatile parts of the human body.

—Nina Jablonski, *Skin: A Natural History*

What is the function of the skin? It is the information source
and processor. It is the foundation for sensory reception. It is
a barrier between us and our environment. It is an immunologic source of hormones for protective cell differentiation. It
protects our underlying organs from radiation and mechani-
cal injury. It serves as a barrier to toxic materials and foreign
organisms. It plays a major role in regulating blood pressure
and the flow of blood. It performs regenerative repair. It works
as a temperature regulator. It is involved in the metabolism and

storage of fat, salt, and water. It serves as a reservoir for food and water and as a respiratory organ for the passage of gases. It is a synthesizer of important compounds including vitamin D. It forms an acidic barrier that protects us against bacteria.

On a morning walk, I stopped to speak with a city worker who was heading into a hole in the ground to service the water pumps. Given my insatiable curiosity, I had to ask him about how the system was hooked up and how our neighborhood water got pumped in. He was informative, with a good tour guide's patience.

"I'm glad you asked," he said. "Most of the time we only get bothered when something goes wrong." He provided me with a mini-lecture about what lies beneath the city surface and how many things must be maintained in order to keep our rising demands for service satisfied.

Underneath our homes, sidewalks, and streets is a rapidly increasing mass of cables, pipes, and devices for the delivery and removal of water, electricity, data, natural gas, and sewage. Most of us have experienced the frustration of having a single one of these malfunction, as when a water line explodes or a utility line is cut.

Likewise, underneath the skin is a complex, intense system—one that outperforms any municipal system. Just as construction workers check and mark carefully before drilling, all of us who penetrate the body's top surface must respect the boundaries of key structures and evaluate the anatomy prior to any invasive procedure. When treating people, we must consider a plethora of factors in order to provide comprehensive care, including personal and family history of disease; current medication use; social history including smoking, alcohol, and education; and the ability to pay for treatment.

The skin is a marvel. In the best circumstances, it heals itself, repairing and restoring its former integrity. It is grim in sorrow, radiates warmth in love, and shines in tranquility. The skin is an organ in and of itself, with its own personality, temperament, and particular eccentricities.

From a very early age, we use one sense the most—touch. When we reach out and touch an object that is too hot, an almost instantaneous event occurs. The temperature sensors in our skin send nerve signals running up the arm to the spinal cord and into the brain at over two hundred miles per hour. The brain interprets the sensation as pain and signals the hand muscles to move away, all before we can think.

Ashley Montagu in his book *Touching: The Human Significance of the Skin* focuses on what he calls the somatopsychic or centripetal approach, the manner in which tactile experience or its lack affects the development of behavior and "the mind of the skin." He writes, "The skin, the flexible, continuous caparison of our bodies, like a cloak covers us all over. It is the oldest and the most sensitive of our organs, our first medium of communication, and our most efficient protector."

The scale of these contributions is staggering. In a piece of skin the size of a quarter, more than three million cells, one hundred to three hundred sweat glands, fifty nerve endings, and three feet of blood vessels are at work. The skin comprises millions of cells of different kinds, some 350 different varieties per square centimeter. Huge numbers of touch, pressure, pain, heat, and cold sensory receptors live on the skin surface, particularly in the hands, face, and mouth. That's why babies like to explore with their tongues: the tongue alone has more than nine thousand sensory receptors.

Montagu writes, "As the most ancient and largest sense organ of the body, the skin enables the organism to learn about its environment. It is the medium, in all its differentiated parts, by which the external world is perceived."

The perception of the skin varies with each culture and time. The skin is considered a mirror of inner wellness in traditional Chinese medicine. In Greek mythology, the skin is a vulnerable shield. The modern, Western view increasingly recognizes skin as a permeable system, and recent research confirms that it has its own built-in, autonomous immune defenses.

We interpret skin as providing evidence for deeper problems, both bodily and psychological. Our skin can torture us by disfiguring our bodies, but it also makes up the most basic facets of our social worlds. Without "normal" skin, we may find it hard to hold a job, feel self-esteem, make friends, or find love.

As Montagu writes, "On our skin, as on a screen, the gamut of life's experiences is projected: emotions surge, sorrows penetrate, and beauty finds its depth. Soft, smooth source of youth's vanity, skin later bears wrinkled witness to the toll of years. Radiant in health, it tingles to the affectionate touch."

If I took all the patients I have seen and put them along an elastic spectrum, I could not stretch it far enough to encompass the contrasts in characters and skin that I have encountered. The differences within this multitude of humanity are as bright as the sunniest day and as dim as the prospects of a few of them rising above the genetic lottery.

During any given week I may see and treat a coal-black Nigerian boy who has barely learned to speak a bit of English; an old, world-worn Southern farmer with tiny islands of coral blue eyes in a sea of sun-damaged skin; a shoulder-stooped, bronzed Cuban man formerly a prisoner under Castro, his face sagging in depression and haunting memories; a homeless skinny black man, washing his feet in the exam room sink, specks of vitiligo everywhere on his skin, as if a dripping paintbrush had splashed randomly onto him; a red-armed backwoods white boy from Arcadia, dragged in by his sleepless mother because his skin has peeled off from eczema and he has missed most of the school year and scratched and hollered every night; a thin, slack-jawed, coffee-colored Haitian woman who arrived in Florida by boat in the middle of the night; and any number of those whose skin cancer or psoriasis or itching has gotten so out of control that they grab onto my office exam table like a buoy in the Atlantic during a hurricane and will not leave until they finally gain relief. In between may be visits with children who have warts or young men and women with acne, who recognize the relatively benign nature of their problems, and may even add some hope or levity to the day.

CHAPTER ONE

WHAT COVERS US?

The finest clothing made is a person's own skin, but,
of course, society demands something more than this.

—Mark Twain

Our skin, at the most basic level, defines us.

The skin is the body's largest organ, averaging twenty
square feet and nine pounds; it makes up 16 percent of the
body's weight. The skin is complicated but amazing in structure; it can be the target and dwelling place of thousands of
tiny viruses, bacteria, or yeasts, yet does a stellar job of living
with the good and keeping out the bad, all to protect the inner
environment.

EPIDERMIS, DERMIS, SUBCUTIS: A TOUR OF THE SKIN

The epidermis, the outermost layer of skin, is a major part of
the immune system. It is filled with Langerhans cells that form
a first line of defense against environmental threats, identifying
foreign materials and dangerous substances and ridding us of

them. The epidermis is also part chemical plant, synthesizing vitamin D in the presence of sunlight, transforming a variety of helpful chemical compounds that interact with it, and inactivating substances that could be dangerous for us to absorb.

The epidermis protects the body from all kinds of insults. As Arthur and Loretta Balin write, "It seems fitting that the organ that defines the boundary between outside and inside worlds would be involved in the essential immune-system task of distinguishing between self and other." Poison ivy is a common example: it is known to affect over 350,000 people in the United States annually, and demonstrates the wide range of possible sensitivities and reactions to exposure. Here's how the reaction to poison ivy works. Urushiol is a chemical within the sap of the poison ivy plant that binds to the skin on contact. Urushiol shimmies its way into the skin and is broken down by T lymphocytes (or T-cells) that recognize it as a foreign substance, or antigen. The T-cells send out inflammatory signals called cytokines and the immune system pumps up the volume and calls in the troops—white blood cells. Still under the command of cytokines, the white blood cells turn into macrophages (super Pac Men) that eat up the foreign urushiol but also cause collateral damage to the normal tissue skin, resulting in inflammation and a dermatitis. A severe allergic reaction and blistering and oozing may occur in susceptible individuals; the fluid is produced by the body as blood vessels develop gaps and leak fluid through the skin. Approximately a quarter of the people exposed to poison ivy have no allergic reaction, while in extreme cases a anaphylactic reaction can occur. With age and repeated exposure, the sensitivity usually decreases.

The epidermis has a top layer called the stratum corneum,

composed of closely packed cells that protect the skin from external abuse. The stratum corneum keeps the skin hydrated, both absorbing water and preventing water evaporation by means of a dense network of the protein keratin. The stratum corneum's thickness varies throughout the body. On the palms of the hands and the soles of the feet, this layer is thicker to provide additional protection. Generally, the stratum corneum has fifteen to twenty layers of dead cells in addition to the layers of proteins, for a total thickness of between ten and forty micrometers—very thin.

Every move you make results in showers of skin particles from the epidermis released into the air. Every twenty-four hours an estimated ten thousand million skin scales or squames peel off each of our bodies, accounting for one to one and a half grams of skin each day, or about one pound each year. These scales are the desiccated remnants of skin cells that continually form at the base of the epidermis and travel slowly toward the surface. After forty to fifty-six days a newly formed cell reaches the surface and it is now called the stratum corneum, the same horny components that make up our hair and fingernails.

At high magnification this surface of dead skin appears as irregular patches of rough and curly cornflakes. House dust consists of 80–90 percent skin; squames are the motes in the sunbeam filtering into our rooms.

Just beneath the stratum corneum live the keratinocytes or squamous cells, which mature and move toward the surface to form the stratum corneum. Below them, in the deepest layer of the epidermis, is the basal layer, containing cells that continually divide and form new keratinocytes, replacing the old ones shed from the skin's surface. This constant upward migration characterizes the epidermis.

Also in the basal layer are the cells known as melanocytes, which determine differences in skin color (more about this in chapter 4). These cells produce the pigment melanin, which protects the skin from sunlight and determines the intensity of skin and hair color. Each of us has the same number of melanocytes. The difference between darker and lighter skin tones is a result of the type, amount, and arrangement of the melanin produced by our melanocytes. Those with darker skin color, such as African Americans, have more melanin and are much better adapted to the harsh conditions of sun exposure. Carotenes, mostly located a level deeper down in the dermis, may contribute to the yellowish cast characteristic of Asian skins. Hemoglobin, the oxygen-carrying pigment in blood, gives pinkness to some fair skins. Freckles are due to increased melanin production, while nevocellular nevi, or moles, are caused by tightly packed groups of melanocytes. Solar lentigines, the flat, brown "liver spots," occur because of an abnormal increase in the number of melanocytes. Those with vitiligo have a decrease in melanocyte function and albinos are genetically unable to produce any melanin at all.

But the epidermis does not only protect and color us. It also is host to some of the marks of aging and sun exposure. If you look closely at the skin of the elderly, you may notice the acquired spots and "barnacles of life" (seborrheic keratosis) that accumulate in the epidermis. Scaly actinic, or solar, keratosis can be detected by feeling its sandpaper texture.

Below the epidermis is the dermis, the middle layer of the skin. It contains blood vessels, lymph vessels, sweat glands, collagen bundles, fibroblasts, and nerves. Held together by a protein called collagen made by the fibroblasts, the dermis is a major contributor to the skin's flexibility and strength and also contains pain and touch receptors.

The dermis also contains the hair follicles; infection or inflammation associated with infection in the vicinity of the roots results in folliculitis. As the Balins have pointed out, we humans call ourselves "naked apes," yet we are covered with fine, unpigmented hairs that are actually ultrasensitive touch sensors. As the only mammals with such highly sensitive touch receptors all over our bodies, we require an enormous brain to process this constant sensory input from the skin.

The dermis can be marked by dilated blood vessels called telangiectasias, or spider veins, brought on by chronic sun exposure, by estrogen, or in some cases by an underlying liver or blood disorder. (As we will see in part 2, every skin disorder has its anatomical correlate; the field of dermatopathology specializes in connecting clinical skin findings to underlying anatomy.)

The subcutis (also known as the subcutaneous layer) is the deepest layer of skin. It consists of a network of collagen and fat cells. This layer helps conserve the body's heat and protects the body from injury by acting as a "shock absorber."

When we are babies, our skin is elastic and resilient—and it becomes less so every day from then on. We lose about 1 percent of our collagen, as well as elastic fibers and blood vessels that attach to the epidermis, every year after age thirty. The result is crinkles and wrinkles, a rather unfair exchange. The skin becomes sallow and pale. We increase fat deposition in the areas we don't like, and we lose fat and therefore insulation in other body areas such as the face, arms, and legs. The underlying tissue depletion makes us more prone to injury, and the loss of nerves decreases our tolerance for cold.

DIAGNOSIS AND ECOLOGICAL DETECTIVE WORK

In examining patients who have skin problems, we note the morphology of individual lesions, their pattern in relation to each other, and their distribution on the body. Since the earliest days of medicine physicians have been observing skin diseases and classifying them by these three criteria. Skin diseases are generally dynamic processes that evolve over their course. Dermatologists often find it helpful to identify "primary lesions," which are the earliest abnormalities, and "secondary lesions," into which they may evolve. Understanding this evolutionary process makes understanding the pathophysiology of the disease possible.

Skin diseases are generally categorized as tumors (abnormal masses of tissue that may be solid or fluid-filled and can be benign, premalignant, or malignant, that is, cancerous), pigmentation abnormalities (such as birthmarks, melasma, vitiligo, and other pigment disorders), papulosquamous diseases, vesiculobullous diseases, papular eruptions, eczematous dermatitis, hypersensitivity reactions, cutaneous infections and infestations, and diseases of the skin appendages (hair, nails, glands, blood vessels).

I greatly respect other medical providers and their tools of diagnosis for inner maladies, from the medical imaging of radiologists to the scans and tests of neurologists. A pathologist can be an enormous aid to a dermatological diagnoses, adding another set of eyes and a deeper view of a disease process caught in a moment in time. Yet perhaps no other field of medicine entertains the notion of visual, real-life pattern recognition more than dermatology. Dermatology lives in the observable and the palpable—the skin. Skin clinicians deal with life in the wild, not tamed and frozen tissue samples from removed body parts and not representations on a screen.

I have always tended to look at the skin from the perspective of habitat and ecology. When I am not seeing patients in my office, I am often in the natural environment of Florida, hiking, kayaking, taking photos, and looking around. I strive to evaluate a skin lesion in the same way. If I see a series of red, dry, scaly actinic keratosis (AKs)—the so-called precancers—I look for the more fully developed cancers. AKs live along the continuum from early stages to more raised and hypertrophic lesions to fully robust creations that manifest as squamous cell carcinomas (SCCs). If I can rid a patient's skin of AKs at an early stage, I eliminate the invasive species before it brings on more damage.

The natural habitat of these AKs and SCCs is on the sun-exposed areas of the skin, particularly on the left arm of those who keep an arm out the window while driving, the face, the neck, the chest, the legs, and any other areas of chronic sun exposure.

All of this comes into perspective when discussing these maladies with my patients. Dermatology skills include the practical evaluation of the topography and climate of the skin. Who hangs out with this particular skin character? What will most likely provide benefit or do any harm? The more deeply I know the disease, the more deeply I can understand the prognosis and potential treatment. Like an ecologist, I have to know where to look if I want to find a familiar species: basal cell cancers (BCCs) and SCCs live in the more superficial skin layers, while melanoma invades subcutaneously. Sweaty armpits (for medical students and fans of Latin, intertriginous areas of the axillae) and upper thighs are delightful arenas for fungi to frolic and breed. The larger environment matters too: to give just one example, the rampant tinea versicolor has been reported to infect up to 20 percent of the population of Florida at any one time.

What would you expect to find as you travel over any particular person's cutaneous terrain? If I encounter the short timber of a cutaneous horn—a hard, horn-shaped tumor—I know to be cautious and to collect a sample that includes the base of the tumor when doing a biopsy, where the squamous cell layer may harbor a cancer. If a patient comes to me with a fire-like eruption on the face, neck, or other areas exposed to sunlight, I consider whether it may signal a photosensitivity reaction based on a medication that makes the skin more vulnerable to the sun's rays. If I see atopic dermatitis (also known as atopic eczema, a rash) on an itchy child, I inquire about the rest of the common triad—allergy and asthma. If I notice a ring of warts on the finger of a seven-year-old, I also look at his or her lips to see if hand-to-mouth behavior has resulted in the spread of the condition via auto-inoculation. If, when I inspect the back of a nervous character, I note that easier-to-reach areas have multiple scratch marks in different stages of healing amid a forest of erratic depigmentation, but that the harder-to-reach center of the back is untouched, I shift my diagnostic weight to self-induced neurodermatitis, pathologic skin picking in the absence of any underlying skin disease.

Some of these skin diseases may have protective utility. Psoriasis, for example, may have a hidden adaptive function that carries a genetic survival advantage. If the same genes that trigger psoriasis also control the intensity of response to bacterial invasion, then perhaps the combined one-two punch of an enhanced inflammatory response and thickened keratin layer have given those with the psoriasis-predisposing genome a survival advantage. The natural process of desquamation, where the skin rids itself of excess layers of keratin, is heightened in psoriasis and may provide a helpful response to discourage colonization of the skin's surface by undesirable microbes and maintain integrity of the skin by shedding faster than colonization can get traction.

Other protective roles for psoriasis can be seen with cutaneous tuberculosis, a disease that can bring on horrible facial destruction. Psoriasis first came to widespread attention in the medical community in the mid-nineteenth century, coincident with a high prevalence of cutaneous (as well as systemic) tuberculosis. As many researchers have reported, cases of patients with both skin diseases were essentially absent. It may be that psoriatic carriers are protected from tuberculosis, or have a survival edge against the more disfiguring cutaneous tuberculosis. If the psoriasis carrier could be protected against tuberculosis, the predisposing psoriasis genotype could survive. Trials and research with the new biologic drugs for psoriasis have proved that psoriasis patients have highly activated immune systems. And one of the main contraindications for the use of any psoriasis-halting biologic is an active systemic tuberculosis infection—in this case, healing patients' psoriasis might actually worsen their tuberculosis.

Dermatology includes pattern recognition as a primary detection tool, and changes in morphology and distribution of lesions on the body are part of each exam. New growths arise and old ones change form, and infections like candidiasis, from a yeast, erupt in locations where and when they find an opportunity for survival. In a similar fashion, you may notice during a walk in the woods a new flower blooming or pay attention to which species of bird frequents a certain oak or elm. Or you may come across a tree down on the side of the trail and decaying, and note that the way the bracket fungi are oriented on the trunk tells you how long the tree has been nonvertical. As with each observation in nature, ecological abnormalities involving changes on the skin due to invasions, disturbances, and imbalances are an integral part of observing skin diseases and providing treatments.

The skin is an amazing, versatile organ and science discovers more about its magic every day. As you read on, you will understand more about the wonders of our covering and what can happen when it gets violated or reflects an underlying problem. You may even pick up a clue or two on how to save your skin.

CHAPTER TWO

CARE AND PROTECTION
OF THE SKIN

Pull down thy vanity, it is not man
Made courage, or made order, or made grace,
Pull down thy vanity, I say pull down.
Learn of the green world what can be thy place
In scaled invention or true artistry,
Pull down thy vanity . . .
The green casque has outdone your elegance.

        Ezra Pound, "Canto LXXXI"

As soon as I walked in the room to introduce myself to Guillermo, I knew he represented a challenge of long duration. Populating his face and neck were at least fifteen easily recognizable skin cancers spread out in odd arrangements and patterns. He had one of the worst collective cases of these cancers that I had ever seen. None of them were immediately life threatening, but the disfigurement was horrible. "I have most of these many years," he said in halting English. "My family tell me I need to get them off."

"Your family is right," I said.

Guillermo spent his first fifty years in Cuba, a country with a poor record of skin cancer prevention. In my own sampling, many of my worst skin cancer patients have been from Cuba. And Guillermo had not fared much better after emigrating, as he had exchanged one tropical country for another. Since arriving in Florida five years ago, he had done little to protect himself from further skin insult and injury.

What's more, our skin cancer prevention program in the United States is not much better than that of Cuba. According to the National Cancer Institute, 40–50 percent of Americans who live to age sixty-five will have skin cancer at least once.

FROM SUN GODS TO TANNING SALONS

Why does our skin get darker over time and exposure to the sun? The main reason is the oxidative stress placed on melanin as it interacts with ultraviolet light, protecting us from dangerous rays.

Melanocytes live in the bottom layer of the epidermis, just above the dermis, and manufacture melanin from an amino acid, tyrosin, with the help of an enzyme, tyrosinase. Exposures of five to ten minutes of sunshine do not bring on tan, but longer than that will initiate a process called melanogenesis. In melanogenesis, ultraviolet (UV) light stimulates the production of melanin in the form of insoluble melanosomes that surround the epidermal cells, which move up to the surface of the skin and result in a tan.

Though people with dark skin and people with light skin have the same number of melanocytes, the way the melanin is distributed and produced is quite different. Darker skin contains more melanin, which protects against ultraviolet radiation and damage to DNA. For this reason it is rare to see skin cancer in African Americans, though I recently saw an African American patient with early skin cancer on his chest. Research by Gloster and Neal, published in 2006, states that the incidence of skin cancer in African Americans is approximately 3 cases per 100,000 people; among European Americans it is 234 cases per 100,000 people. For both African Americans and European Americans, the primary risk factor is chronic UVB exposure.

Many societies today value the look of a dark tan, causing many people to expose themselves to high levels of UV radiation. But no safe tan exists; all tanning is another form of burning.

A quote that has been erroneously attributed to Kurt Vonnegut, but in fact was written by newspaper columnist Mary Schmidt, captures some of the importance of protecting skin from the sun. According to the urban legend, Vonnegut was asked to be the guest speaker for a prestigious college's commencement exercise, and told the graduating class only: "Wear sunscreen. If I could offer you only one tip for the future, sunscreen would be it. The long-term benefits of sunscreen have been proved by scientists, whereas the rest of my advice has no basis more reliable than my own meandering experience. . . . Advice is a form of nostalgia. Dispensing it is a way of fishing the past from the disposal, wiping it off, painting over the ugly parts and recycling it for more than it's worth. But trust me on the sunscreen."

During my childhood summers in Michigan, where I grew up less than an hour from the beach, I cherished the sun. I recall the smell of baby oil mixed with the thin toasty smell of heated skin. I savored the precious two months when I could actually lay in the sun with scant clothing, the memories of giant snow piles melting to my mind's periphery. As an adult, I began to consider that what I had thought of as a benevolent sun was instead some type of malicious nuclear reactor that had turned my DNA into a cancer-making machine. I added eternal vigilance about the devastating effects of the sun to all the other problems of day-to-day life—terrorism, Lyme disease, global warming, and long lines at Starbucks. And, as a dermatologist, I help others heed the wake-up call.

The healing power of the sun has always been evident. The first Greek sun god was Helios—his name gives us the term dermatoheliosis, for photoaging or sun damage on the skin. Later Apollo, also the god of health and prophecy, became the accepted sun god. Apollo's mortal son Aesculapius was said to be the first physician, and his staff entwined by a serpent is Western medicine's symbol. Where science is concerned, we have known for over five hundred years, since Copernicus asserted it, that the sun is literally the center of our universe. From cave dwellers to the present day, humans have borne witness to the sun's healthful properties: its germicidal powers, its ability to diminish various skin diseases such as psoriasis, its vitamin D synthesis, and its feel-good effects (an antidote to seasonal affective disorder for those in dreary northern climates), to name a few. But perhaps not even Apollo could foresee the sun's damaging effects on future generations.

How did this oceanic change in our sense of the sun come about? Much of the history of suntanning carries with it a media hype akin to that expended on cigarette smoking. During the 1920s and 1930s, many movie stars were paid to smoke at attention-getting locations, enabling tobacco companies to increase their sales dramatically. In similar fashion, the Coco Chanels of the world launched huge tanning spikes. Betty Grable, Rita Hayworth, and other bathing beauties were pictured in one- and two-piece bathing suits exposing their tan bodies.

European women sunbathed in decorative, attention-getting sun hats and shawls for fashionable reasons, not protection. And if their skin happened to have an untanned spot, tinted powders and creams were available. The fashion world created shoes to be worn without stockings and sleeveless dresses for women wishing to expose their tans. Although ivory skin had once been associated with wealth (and not working outdoors), this shifted in the social upheaval that followed World War I. A tan in the winter was a clear sign that the tan bearer had enough wealth and leisure to afford an exotic, warm climate.

In 1929 Helena Rubinstein, the cosmetics magnate, warned, "Sunburn menaces your beauty." But women's magazines pushed tanning. Cosmetic companies introduced suntanning oils; the ingredient PABA was introduced in 1943. While public ambiguity about the look persisted, women's magazines encouraged sun lamps and suntanning and cosmetic companies introduced suntanning oils. The first so-called self-tanning product, Man-Tan, a tinted lotion that created the effect of tanning, hit the market in the mid-1950s, with beige, brown, or orange results. Certain science reporters used women's magazines to suggest that gradual tanning could actually cancel out the sun's damaging effects. The media slowly took in the message of the sun and the skin. *Harper's Bazaar* reported in 1954, "There are sunscreen preparations that can cut the intensity of the sun's rays by 75 percent." In my medical practice, there is not a week that goes by when someone does not say to me, "We didn't know about the problem with the sun when we were younger." They may not have known about it, but it was already evident, if sometimes obscured, in the media.

Meanwhile, dark skin became a status symbol in the 1960s. Suntanning was no longer a spectator sport, especially among the young. Coppertone advertisements filled the airways— "Tan, don't burn, get a Coppertone tan!"—and beach movies filled with bikini-clad teens populated the television. In the 1970s, with gallons of Johnson's Baby Oil coating many an unsuspecting epidermis, another industry began to blossom— the indoor tanning industry. Even those in the cold North could try to keep a tan or could prepare for adventures to warmer climates with a series of trips to the tanning salon. Tan skin was a sign of having leisure time. The Hawaiian Tropic TV ad featured a beautiful blonde who sensually said, "White is for laundry."

In the early 1970s, the Food and Drug Administration (FDA) began to treat sunscreens as over-the-counter drugs and not cosmetics. More stringent labeling was required. After the FDA began regulating sunscreens, the makers of Johnson's Baby Oil warned that the heroine of their ads should "take a little less sun." In 1978, the FDA declared sunscreens to be safe, effective, and useful to prevent skin cancer and sunburn and to slow premature aging of the skin. The SPF (sun protection factor) numbering system was developed using numbers two through fifteen.

During this period, the tanning industry, almost entirely unregulated, continued to prosper. Home-tanning units were particularly damaging: in addition to their cancer-producing and premature wrinkling effects, they emitted high levels of UVB light that burned the skin and didn't tan. More advanced tanning units emitted both UVB (ultraviolet B or shortwave rays) and UVA (ultraviolet A or long-wave rays) and brought on further skin damage. UVA rays account for up to 95 percent of the UV radiation reaching the Earth's surface and are less intense than UVB, but penetrate deeper. Tanning booths primarily emit UVA, using high-pressure sunlamps in doses of UVA as much as twelve times that of the sun. Ultraviolet (UV) radiation is responsible for 90 percent of the visible signs of aging on the skin of fair-skinned people.

By the mid-1980s, public education programs about the dangers of overexposure to the sun and the problems with tanning began to grow. The American Academy of Dermatology voiced its strong support for sun protection, and sunscreen manufacturers produced higher SPF products.

But the damage was done. Studies showed a 500 percent increase in malignant melanoma incidence between 1950 and 1985. A 1987 American Academy of Dermatology study revealed that 96 percent of Americans knew that the sun caused skin cancer. One-third of these same adults admitted that they had deliberately worked on a tan.

In the 1990s, the indoor tanning industry continued to be one of the fastest-growing businesses in the United States. The average age of indoor tanning patrons was twenty-six, mostly women. Almost two million of these patrons were considered "tanning junkies," who made almost one hundred tanning parlor visits each per year. In 1991, eighteen hundred injuries were reported from tanning devices.

The pushback against tanning continued. Fashion design industry leaders such as Eileen Ford stated, "The tanned look is dead." The American Academy of Dermatology stated there was "no safe way to tan," following a consensus conference on photoaging and photo damage. And natural-looking tans sans streaking or discoloration became possible with improved tanning products. Wide ranges of protection against UVA as well as UVB radiation were created by the sunscreen industry in 1990 in response to the rising tide of information about skin cancer facts—600,000 new cases of skin cancers, 6,300 deaths from melanoma, and 2,500 deaths from squamous cell carcinoma in the previous year. The role of genetics, ozone depletion, and other skin cancer production factors took center stage.

The incidence of skin cancer continued to increase, with 700,000 new cases of skin cancer diagnosed in 1993, 32,000 of them malignant melanoma. Yet the popularity of tanning refused to wane. In a 1996 survey of young adults, 58 percent confessed to having at one point worked on a tan and 62 percent thought that people look better with a tan. In 1997, two-thirds of teens surveyed in *Seventeen* felt they looked better and felt healthier and more sophisticated with a tan. Half of them stated they looked more athletic with a tan. The tanning industry grew to almost twenty thousand salons in the United States, serving twenty-two million customers per year; tanning salons were regulated in only half of the states. The American Academy of Dermatology continued to warn the public to minimize the sun's damage to the skin and eyes by planning outdoor activities to avoid the sun's strongest rays, wearing protective covering, wearing sunglasses, and always wearing a broad-spectrum sunscreen.

SEEING THE LIGHT ON TANNING

Why do so many people pound their skin with artificial rays to get a tan? With so many studies linking UV radiation with skin cancer risk, it might be a good time to close the doors of the fake-n-bake. According to studies from the American Academy of Dermatology and the U.S. Department of Health and Human Services, tanning increases the risk of melanoma, squamous cell carcinoma, and basal cell carcinoma; and excessive exposure to UV radiation during indoor tanning leads to skin aging, immune suppression, and eye damage, including cataracts and ocular melanoma. Yet today more than one million people tan in tanning salons on an average day, nearly 70 percent of them girls and women aged sixteen to twenty-nine.

And tanning continues to be big business. According to a new study by San Diego State University public health researchers, indoor tanning salons in America's big cities often outnumber Starbucks or McDonald's. The study looked at the number and density of indoor tanning facilities in 116 of the largest cities in the country and then compared tanning facilities with the two above-mentioned "ubiquitous institutions." Charleston, West Virginia, took the top prize for the highest density of tanning salons, with a total of eighteen facilities for a population of 53,000, with only one Starbucks and seven McDonald's. Indoor tanning megacities included Pittsburgh, Portland (Maine), Akron, and Columbia, South Carolina.

Interviews with six thousand teens across the country, included in the San Diego State University study, revealed that proximity appears to trump rationality: living within two miles of a tanning salon increases usage. Sales incentives like monthly discount packages and extended hours also bump up the numbers. According to a *60 Minutes* report, in 2013 Americans spent 6.6 billion dollars on sun care products.

Why are so many people still tanning, especially with all the evidence regarding the cumulative effects of UV radiation? Why do we have this mad desire to feed the melanocyte? There are nearly as many answers as there are tanning parlors.

Some people say they do it for the vitamin D. But it is safer to eat cheese, drink milk, eat fish, or take vitamin supplements to replenish vitamin D. The relaxation or pleasurable feelings that some report from tanning may even be addictive, but exercise is a healthier alternative. For those who suffer from Seasonal Affective Disorder, light boxes can bring a smile, even on cloudy winter days, and use healthier broad spectrum light, not UV. Others have promoted "good UVA rays" as opposed to "bad UVB rays," but all data show an increase in skin cancer and skin aging with UVA rays from tanning. Others suggest that tanning helps with skin problems such as psoriasis, but anyone with that concern should seek environments such as a doctor's office, where the type and dosage of light is regulated. As for those who feel "healthier" with a bronze tint to their epidermis, there are safe self-tanning lotions and spray-on tans.

Tan-seeking buyer beware! The Obama administration, to help fund the 940 billion-dollar health-care overhaul, added a 10 percent tax on receiving indoor tanning services: the initiative is expected to generate 2.7 billion dollars over ten years.

I know it is difficult for people to change their behavior. Knowing that tanning increases the risk of skin cancer by 75 percent is important—we know that tanning beds are just as much a carcinogen as cigarettes. But the habit is hard to drop. For one, light is also known to increase endorphins, and not everyone wants to run a mile to get the same effect. Just as there are patches and pills and other products designed to help people to quit smoking, SPF 15 and higher sunscreens have been readily available since the early 1980s in a variety of topical preparations, for those who want to be in the sun more safely. I certainly preach the gospel of using sunscreens, especially to the parents of vulnerable children, as sunscreen use will pay off in huge savings and in treatment prevention. I regularly see patients in their twenties and thirties with skin cancers and have had both a fourteen-year-old and a fifteen-year-old in my office presenting with stage IV malignant melanoma.

The waves of history lap onto the shores of today. The unprotected sun exposures of all these earlier generations are now causing a bloom of skin cancers. As a dermatologist I use whatever mechanisms are necessary to rid my patients of skin cancers at all stages of development.

Those of us working in dermatology sprint around looking for ways to prevent and treat skin cancer using chemoprevention, radiation, laser, cryotherapy, and surgical excisions. We are the gardeners of the skin, detecting and removing cancers that sprout up like unrestricted weeds. We can only get rid of what we can see, and, therefore, we advocate the importance of skin cancer prevention, detection, and treatment.

.   .   .

What were my options with Guillermo? He had the wrong kind of insurance to allow for advanced surgeries and not enough money to pay for anything cosmetic. With his consent I chose to perform a series of shave biopsies and electrodessications and to allow his healing skin to do the rest. "I am very happy to get better," he said. Over time he improved, exchanging skin cancers for scars, a trade-off that occurs thousands of times a day throughout the world that Apollo once ruled.

CHAPTER THREE

THE HIDDEN LIFE ON THE SKIN

Adam
Had 'em.

—Strickland Gillilan, "On the Antiquity of Microbes"

Inspecting our front yard sago palm on a clear fall day, I noted a thin coating that looked like frost, except this was Tampa, and the weather was eighty degrees. On a closer look, the sparkling sun revealed an infestation of Asian cycad scale *(Aulacaspis yasumatsui)*, a blight that sucks the life out of king and queen sago palm trees. The microscopically tiny white insect has an armored coating and multiplies faster than a duck crossing a superhighway with its tail on fire, reaching a density of about three thousand per square inch. In a few months, a fifteen-foot queen sago with a spread of over twenty feet will be covered, stems and leaves, with a scale of the nasty beasts. The critters are difficult to eradicate. Even though they don't fly, they can become airborne: when there is a strong wind, the immature form of the insect can lift off and land on a sago up to a half mile away. If left uncontrolled, they will eventually kill the plant. I searched the Internet and found a few helpful tips to save my palm friend from certain demise.

I have treated tens of thousands of my fellow Homo sapiens for skin diseases brought on by fungal, bacterial, and viral invaders. Questions like the ones I asked about my sago palm are relevant for us too. What inhabits us? How many of the flora that populate our skin are normal inhabitants and what portion is potentially hazardous?

One such patient was Theresa, a feisty twenty-six-year-old who worked at one of the local "exotic dance" clubs, a major industry on the West coast of Florida, along with churches, beach resorts, and other jobs tied in with skin and salvation. I had treated a friend of hers for another problem with success. With her friend's recommendation, Theresa showed up on my office doorstep one day, desperate to get rid of a "bad rash." Her boss had told her to get it taken care of as soon as possible and she wanted to get back to work.

"This rash shows up under the lights at work," she told me and my medical student, an eager young trainee named Jackie. Theresa lifted her blouse to show the rash covering her upper back. "I asked some of the other dancers to give me some ideas to get rid of it. But it's dirty in that place and I probably caught it there." No matter what Theresa and her coworkers dreamed up to remedy the situation, the rash did not go away. When Theresa had finished explaining, I shut the office door, turned off the lights, and shined a handheld device called a Wood's lamp on Theresa's back: the fungus fluoresced, indicating a fungal skin disease called *Tinea versicolor*. The disease *Tinea versicolor* is not contagious and is one of the most common skin diseases in tropical and subtropical areas of the world. The yeast (yeast is a type of fungus) that causes *Tinea versicolor* lives on everyone's skin, and it's not clear why the yeast overgrows on some people's skin and not that of others.

I made sure that Jackie not only saw the glow, but also touched the rash. One of my goals as a clinical instructor is to guide students who are dermatologically visually illiterate to see the important patterns of the more common skin diseases. I also want them to be able to feel for the subtle changes that occur on the skin surface. Sometimes I must hold the hands of timid students as they both feel the disease with their fingers and take a visual snapshot, so that both senses together etch a complete picture into their minds. When this picture is combined with a patient's story, it affords an accessible grasp of many diseases—a grasp more accessible in dermatology than in any other diagnostic discipline.

"That's the same thing I see at work under the lights!" Theresa exclaimed as she looked at herself and the yellow-green appearance of the rash under the lamp.

After reviewing her medical history, I prescribed her a combination of treatments—pills, cream, and shampoo—for as rapid a relief as possible.

"Thanks so much," she said.

"Be patient," I said. "It takes time for it to go away."

TINY WORLDS: FLORA AND FAUNA ON OUR SKIN

We humans are simply one of over two million species of animals and plants. And like our fellow mortal inhabitants, we are at the mercy of tiny viruses, bacterium, and yeasts. Each of us has about as many bacteria and yeasts on the surface of his or her skin as there are people on earth. The life that lives upon us not only puts our lives in perspective but also allows us a peek at a world within the worlds of our integument, a view that is both strange and amazing to fathom.

Excuse the pun, but it is all a question of scale. Fleas have parasites. Bacteria can parasitize the parasites of fleas. Viruses can parasitize the same bacteria. As Jonathan Swift put it:

> So naturalists observe, a flea
> Hath smaller fleas that on him prey:
> And these have smaller fleas to bite 'em,
> And so proceed ad infinitum

Viruses are the smallest live inhabitants of our skins. A virus can only reproduce by entering a living cell and fooling it into making more of the virus's own genetic material. The viruses multiply inside the captive cell until it bursts, releasing more virus to colonize other cells. When we have any lowering of our resistance—an infection, sunburn, or stress—the herpes simplex virus, which brings on "cold sores," may step into the picture. The virus, which usually begins in childhood, may appear on the skin and then return to the underlying nerves, ebbing and flowing based on the individual's immunity. Once infected, the virus is carried for life; more than 90 percent of the population carry the herpes virus.

Yeasts, like Theresa's *Tinea versicolor,* and fungi are inhabitants of the human skin. A yeast is a single-celled fungus that reproduces by budding. The daughter cell grows out from the parent and eventually breaks free. *Pityrosporum,* which belongs to the family Cryptococcaceae, is the most common genus of yeast on our skins. Each organism of *Pityrosporum ovale* is an oval sphere about two microns wide and four microns long; they flourish on our hair and on the fatty parts of our skin. On the scalp and around the nose the *Pityrosporum ovale* population can total half a million per square centimeter. Its cousin, *Pityrosporum orbiculare*, a round yeast about two microns across, can bring on problems when its hyphae expand, forming a spreading mycelium or root-like growth.

I have often seen patients with yeast infections following the chronic use of steroid creams and ointments. Steroids are helpful to calm the inflammation of diseases such as eczema, but they also can suppress the body's natural immune defenses, which sets up yeast for a sumptuous feast. Normally most of the microscopic flora are in balance, but when the protective flora are decreased and the person is immunocompromised, others may take the opportunity, and an opportunistic infection occurs. Many women get vaginal candidiasis and discharge when on antibiotics due to this change in equilibrium.

Of course, it's not just flora, like yeasts and fungi, that dwell on our skin. Our skin is also host to fauna, its own tiny wildlife. When you think of mites you may think of the disease-filled Middle Ages. But *Demodex* is as jovial and well adjusted, if I may be anthropomorphic, on clean hair as on dirty and craves blue blood as much as red. The parasites of the human body, in fact, have shown no respect for social order or class as they have evolved with us through the millennia.

Demodex mites are part of normal human fauna, our constant miniature companion throughout life. These mites, each one-third of a millimeter long, are in the class Arachnida, along with ticks, spiders, and scorpions. *Demodex* mites are common feeders on the hair and sebaceous glands in mammals, benefiting themselves without harming the host; *Demodex folliculorum* usually dwells on the face, and *Demodex brevis* commonly infests the chest and back.

However, there is no consensus to what degree the mites are causative of the skin pathology and how they might contribute to the disease. Although the individual movements of each mite are below the threshold of sensory perception, the effect of their presence is still in dispute. As many as twenty-five mites have been found hanging on to one human eyelash root, a density that calls into question their benignity.

We do know that in immunocompromised hosts, *Demodex* may overpopulate and bring on a dermatitis, an inflammation of the skin that may elicit an itchy rash on swollen, reddened skin. We also know that *Demodex* mites are implicated in demodectic alopecia or "human mange"; the related mite that causes mange in dogs looks identical but is unable to live on humans. *Demodox* may also be involved in the inflammatory reaction in rosacea, a multiphasic disease associated with redness and pustules .

As Michael Andrew writes of *Demodex* in his 1976 book *The Life That Lives on Man,* "Nothing amongst all the unsuspected secrets of one's skin is more astonishing than the thought that the roots of one's eyelashes are colonized by mites. Few people can confront with equanimity the idea that worm-like creatures which have been likened to eight-legged crocodiles squirm out their diminutive lives in warm oily lairs in our hair follicles."

DEFENDERS AND INVADERS:
LEARNING TO LIVE WITH BACTERIA

What else of our normal flora? The skin is sterile at birth but only remains so briefly. Examining the umbilicus for *Staphylococcus aureus*, a common bacterium, shows 25 percent colonization in the first day of life with a steady increase from then on. We have two types of normal skin flora—transient and resident. Resident flora are capable of multiplication and survival and are found as the dominant component in most skin areas. Resident populations on our skins and on the sweeping bristles (cilia) in our air passages generally protect us from the incursions of foreign organisms.

Resident flora include *Propionibacterium acnes*, a prototype anaerobic diptheroid, found in large numbers near the sebaceous follicles of the skin in moist areas. These organisms may contribute to the inflammatory component of acne. *Corynebacteria*, another genus of resident flora, also reach their maximum density in high moisture areas, and like to congregate in the underarm area and between the toes. When any skin bacteria break down the natural secretions from the sebaceous, sweat, and apocrine glands, body odor occurs. Washing with soap and water helps.

In contrast to *Staphylococcus aureus*, which is found on only 20 percent of people, *Staphylococcus epidermis* is uniformly present on normal skin. The huge numbers of this resident flora exert a suppressive effect on other organisms wishing to colonize. Anaerobic staphylococci are also constantly present. However, they have population densities well below those of other resident flora and unlike other staphylococci do not increase in numbers in the case of dermatologic disease. Gram-negative organisms (bacteria that do not retain the crystal violet dye in the gram staining procedure and are more resistant against antibiotics due to their relatively impermeable cell wall) such as *Escherichia coli,* various *Proteus* bacilli, and members of genus *Enterobacter* are uncommon on normal human skin except in moist areas where skin touches skin, such as toe webs, underarms, and groins.

In contrast to these resident flora, transient skin flora are microorganisms—including bacteria, fungi, and viruses—that temporarily colonize the skin by direct skin-to-skin contact or indirectly, via touching objects. That's why it's important to wash your hands, which deactivates the microorganisms of the transient flora. Aerobic bacteria that form spores, such as *Bacilli,* various streptococci, and members of genus *Neisseria,* may briefly visit. Specific ecologic data on these bacteria are difficult to obtain due to sampling and the transitory changes that occur in each part of our skin.

Both resident and transient flora are sometimes known as opportunistic pathogens, generally nonpathogenic bacteria and fungi that can trigger infections in debilitated or compromised hosts. In conditions when the skin is immunocompromised, such as in severe eczema, secondary infection by *Staphylococcus aureus* is common.

In fact, skin disease due to *Staphylococcus aureus* is the most common of all bacterial infections. Impetigo, with its characteristic yellow crusts and blisters, and folliculitis, a circumscribed infectious process originating in a hair follicle, are generally the most superficial of all staphylococcal skin infections. In folliculitis, tiny red pustules congregate around hair follicles. When the infection is recurrent and chronic in the beard area it is called sycosis barbae. Folliculitis may in turn develop into furunculosis, or boils, deep-seated inflammation around a hair follicle. More than 1.5 million cases of boils occur annually in the United States alone.

Methicillin-resistant *Staphylococcus aureus* (MRSA) is a strain of bacteria that is part of the normal flora of the skin and nostrils of many people. But if the MRSA bacteria colonize those body parts and crowd out other normal flora, infections can occur. Although MRSA infections occur primarily on the skin, they can also occur in the lungs (causing pneumonia) or the blood (causing sepsis). Since the MRSA bacteria are resistant to most antibiotics normally used to treat staphylococcal infections, these infections are serious and difficult to treat, particularly for those in a weakened state.

Our tissues are particularly vulnerable to infection in the operating theater, especially in those patients undergoing extended surgeries such as hip replacements. *Staphylococcus aureus* and other opportunistic bacteria can flourish in a wound, creating an invasion of the patient's own bacteria. However, the primary concern in hospitals is cross-infection by resistant organisms.

Although Leeuwenhoek discovered bacteria, it was two centuries later when Pasteur tied in their existence to disease in Homo sapiens. Following his discovery, Pasteur suffered from a morbid fear of dirt and infections; he avoided shaking hands for fear of contamination.

Pasteur's connection between bacteria and disease helped to create our world, in which cleanliness is practiced with religious zeal. Our television screens incessantly proselytize, at enormous costs, for the death of germs and their by-products via disinfectants, deodorants, sprays, and cleaning chemicals. But perhaps with a little knowledge, we may find that the presence of germs on skin is not so terrible.

Although it's tempting to envision life on the skin as the creeping and hopping evident in larger creatures, the huge majority of the inhabitants of our own private zoological gardens are harmless or beneficial. Each of us supports billions of creatures; since no one of us can escape from our animal origins, it is wise to understand what is happening. Just as we have only begun to explore the undersea world and outer space, the world of our skin is still a great mystery. Our skin is an ecosystem, just like rivers and forests, and carries with it all the same issues—self-sustaining boundaries, competition for food and growth, and intimate interconnections between the underlying geography and resident and transient flora. When a person takes a broad-spectrum antibiotic, such as tetracycline, he or she does so with the risk that the diverse set of microclimates of our ecosystem will suffer from an imbalance. Our skin has no seasons or diurnal variation and comparatively limited temperature ranges, but it has the same complexity and need for ecological integrity as many ecosystems. And given adequate nutrition and care, the skin has tremendous self-healing power.

Michael Andrews wrote in his book *The Life That Lives on Man,* "Wherever man goes he is not alone. Though we may leave the Earth we take with us on any voyage of discovery our own personal world that is yet to be completely explored. We evolved on Earth, but we did so in the company of the minute creatures that live out their lives on our bodies. We should treat our fellow travelers with respect; they are much more adaptable than we are, and they do us more good than harm."

Perhaps the old adage that "what you don't see can't hurt you" applies. The huge majority of those critters that live on the skin are invisible and merit our indifference. Luckily, with a few exceptions, we can also fight the critters that float through the air and land on our skin. Ashley Montagu wrote about the skin, "Like a frontier of civilization, it is a bastion, a place at which skirmishes are fought and invaders resisted, our first and final line of defense." And when invasions do bother us, at least we have treatments. As far as I know, we are the only species to have dermatologists, nail salons, beauty parlors, and myriad other means to rid our body of imperfections, real or perceived, and generally improve our appearance in culturally determined ways. I am forever humbled, for along with my fellow soldiers who fight these everlasting skin diseases, I know we can never win the battle. However, I still choose to fight, to provide momentary solace from the onslaught of our own invaders. And when I think about the scale-infested sago, and the thousands of patients, like Theresa, I treat for skin infestations every year, I'm glad I can provide a little help along the way.

CHAPTER FOUR

MELANOCYTES AND
THE COLOR OF HUMAN LIFE

Skin pigmentation provides one of the best examples
of evolution by natural selection acting on the human
body.

—Nina Jablonski, *Skin: A Natural History*

A patient of mine named Sophia brought her fourteen-year-old son, Anthony, to see me. "He's getting some white spots on his face and his hands," she said. "They started off small but now I can see they have gotten bigger." As he sat on the examining table, Anthony appeared as placid about his skin as his mother was agitated. Sophia pointed out the discrepancies in skin color that had appeared over the last several months. Given the emotions that can come up for a parent in this situation, it was understandable that she would need reassurance. I explained to them that I would need to take a small skin sample to make the diagnosis. "Whatever the results, we have seen it before and we can help," I said.

She showed me a medicine that her family had used in Cuba for a similar problem. It is made of human placenta and I had seen similar versions many times before. All are intended to stimulate melanin production, but I had not observed much response.

My main concern with Anthony and also the underlying concern of most parents in such cases is to check for vitiligo. The diagnosis requires a biopsy and often a check for any systemic problems that may be part of the etiology.

In Anthony's case, the biopsy did reveal vitiligo, and he had no underlying abnormalities that could be detected as contributing to the problem. Treatment options vary according to the extent and location of the disease's progression. For Anthony, I prescribed a topical cream. He also began treatments with our excimer laser system, which delivers a high dose of monochromatic UVB right on the under-pigmented skin to bring the color back while sparing the normal skin from harm.

Vitiligo (also known as leucoderma) is caused by the absence of our main characters in this chapter—the melanocytes—and results in acquired hypopigmentation, or loss of skin color. A family history of the disease exists in at least 30 percent of vitiligo patients, and both sexes are affected equally. About half the people who develop this skin disorder experience some pigment loss before the age of twenty. Even though most people with vitiligo are in good general health, they face a greater risk of having increased or decreased thyroid function (hyperthyroidism or hypothyroidism), vitamin B12 deficiency (pernicious anemia), decreased adrenal function (Addison's disease), round patches of hair loss (alopecia areata), and/or inflammation of the eyes (uveitis).

Vitiligo affects 1–4 percent of the world population, and since ancient times patients with vitiligo have suffered the same social abuses as lepers. In fact, vitiligo was referred to as *shweta kustha* in India, meaning "white leprosy." Although vitiligo is disfiguring to all patients, it is particularly so in dark-skinned people, due to the strong contrast.

Davinder Parsad and other researchers carried out several research projects on the psychosocial implications of pigmentary disorders in Asia. In India, those with vitiligo face severe psychological and social problems. According to certain Indian religious texts regarding reincarnation, a person with vitiligo in this life did *guru droh*, or sinned, in a previous life. Vitiligo is particularly debilitating among young unmarried women, who have a slim opportunity for arranged marriage due to the disease. If a married women develops vitiligo, the marriage may end in divorce. According to a recent article, the first prime minister of India, Jawaharlal Nehru, once ranked vitiligo as one of three major medical problems of India, behind leprosy and malaria.

ALBINISM: LIVING WITH DIFFERENCE AND DANGER

Another of my patients, Frances, had a helpless look on her face as she sat down to be examined. "I've got spots all over," she said.

I took a closer look at her fair skin and white hair. On the sun-exposed areas of her arms, face, and neck, red and irritated precancerous and cancerous skin took up more space than any areas of normal skin.

"I just got insurance. Ain't had any insurance in more than ten years. Reckon I better get in and see if a skin doctor could help me."

As we talked, the problem lurking behind her skin cancers started to appear in focus for me. She wore sunglasses. When she removed them, she showed signs of nystagmus, an irregular rapid movement of the eyes back and forth, and strabismus, the muscle imbalance of the eyes also known as crossed eyes or lazy eye. The sunglasses reflected her sensitivity to bright light and glare, called photophobia.

"I can't see too good," she said.

The clues—her sea of cancerous skin on a bed of white, her white hair, her light-sensitive eyes with involuntary movements—came together for me. Frances had albinism.. Her chief diagnosis was oculocutaneous albinism, which carries a number of signs and symptoms, including extremely poor visual acuity: most people with this diagnosis are legally blind. All people with albinism are sensitive to glare, but they do not prefer darkness, and they need light to see just like anyone else.

She had developed a huge basal cell cancer on her right temple. I removed a piece for pathological inspection.

I asked her if she knew much about her albinism. "Somebody once told me something about it, but I can't remember much."

What is this disease?

The word *albinism* refers to a group of inherited conditions. People with albinism have little or no pigment in their eyes, skin, or hair, because their melanocytes do not make the usual amounts of melanin. With the exception of one type of ocular albinism, which is passed on from mothers to their sons, inheriting albinism requires albinism genes from both parents.

One person in seventeen thousand in the United States has some type of albinism. Albinism affects people from all races. Most children with albinism are born to parents who have normal hair and eye color for their ethnic backgrounds. Often people do not recognize that they have albinism.

What do people with albinism look like? Most people with albinism look like Frances, with very light skin and hair. In less pigmented types of oculocutaneous albinism, the type of albinism that affects both the skin and the eyes, hair and skin are cream colored. In types with slight pigmentation, hair appears more yellow or red tinged. People with ocular albinism (albinism that only affects the eyes) usually have normal or only slightly lighter than normal skin color.

A common myth is that people with albinism have red eyes. Different types of albinism exist, and the amount of pigment in the eyes varies. Although some individuals with albinism have reddish or violet eyes, most, like Frances, have blue eyes. Others have hazel or brown eyes.

I have had other patients with this disease. One was a man named Moses, who had grown up in South Africa. "I'm new to this country," Moses said. "Things were not good where I came from. Because of my skin, people there thought I was evil." We had discussed his history and he revealed he had been kept out of school and almost totally ostracized in his community because of his condition. And, like Frances, he was afflicted with skin cancers.

. . .

The striking appearance of albinism has fascinated human-kind for centuries, drawing reactions ranging from veneration to alienation. Albinism was noted in the earliest medical literature. Several ancient Roman authors (including Pliny the Elder, Publius Tacitus, and Aulus Gellius) described albinism in humans. The enzyme tyrosinase is required to make melanin in the skin, hair, or eyes, and albinism reflects a tyrosinase deficiency. Albinism in animals was first demonstrated in 1904, and the first accurate scientific paper written about albinism was by Sir Archibald Garrod in 1908.

In some Asian societies dating back to ancient times, and in Europe during the Middle Ages and the Renaissance, fair skin was considered very attractive, a sign of wealth and high social status. Tanned skin meant that one was obligated to work outside for one's livelihood. By the eighteenth century, the powdered white wig worn by wealthy Americans and Europeans was taken as a sign of the wearer's ability to afford luxury items and be identified as one of the educated elite.

Nevertheless, in nineteenth-century America, albinism—a disease that produces naturally white or pale skin and hair—was considered such a bizarre trait that people with this condition were exhibited in circus sideshows. Furthermore, with the advent of the camera, these individuals were featured on postcards, which were widely distributed and collected from the 1870s to the 1890s. Photo studios such as those of Charles Eisenmann, Obermuller and Son, and Matthew Brady used the photographic art to take pictures of what were regarded as human oddities.

Traditionally, many Native American and South Pacific tribes believed that human beings and animals with albinism were messengers from divine entities. Some saw them as good omens and treated them with great respect. Others viewed their presence as a manifestation of wrongdoings within the community.

In many parts of Africa, including Tanzania, Kenya, the Democratic Republic of Congo, Burundi, and Swaziland, life is particularly difficult for people with albinism. In these regions, widespread poverty and ignorance about the condition deprives individuals with albinism of much-needed protection from the sun. As a result, many die prematurely from skin cancer. Even if they do manage to avoid strong sunlight, it often means a life of virtual solitary confinement and prohibition from participating in the daily activities of their communities.

Tanzania in particular is a nightmare for those with albinism. Albinism is roughly ten times more common in Tanzania than in the world as a whole. Because of bizarre beliefs by certain witch doctors, who far outnumber conventional doctors, people with albinism live in constant terror. More than seventy people have been killed in the last five years by those hunting albinos for their legs, arms, hands, and blood, which are put into potions to supposedly bring good luck and wealth. Those who survive these horrendous acts of brutality can be left with double amputations, their limbs cut off and sold in the underground economy. A 2008 BBC report noted that one witch doctor had offered two thousand dollars for the arm of an albino.

Even today, a plethora of misconceptions about albinism persist. Bizarre characters labeled "albinos," with snow-white skin and hair, blood-red eyes, and supernatural (usually villainous) powers plague the entertainment industry. In parts of Africa, people with albinism have been institutionalized and/or stripped of educational and vocational opportunities due to a misguided belief that the poor vision accompanying the condition (on average, 20/200) prevents one from being able to function adequately in and contribute to society. In these same regions, some members of the medical profession have even been known to recommend abortions to mothers carrying fetuses with albinism because it was thought that such children would die early and would fail to lead productive lives.

In the United States, people with albinism live normal life spans and have the same types of general medical problems as the rest of the population. Those who do not use skin protection may develop life-threatening skin cancers. If they do use appropriate skin protection, such as sunscreen lotions rated SPF 20 or higher, and opaque clothing, people with albinism can enjoy outdoor activities even in summer.

People with albinism are at risk of isolation, because the condition is often misunderstood. Social stigmatization can occur, especially within communities of color, where the race or paternity of a person with albinism may be questioned. Families and schools must make an effort not to exclude children with albinism from group activities.

Because Frances's condition had been untreated for so long, I asked her to return multiple times for treatment. On her second visit, I did a more aggressive surgery and removed the remainder of her facial skin cancer. Over the following weeks

I performed a variety of surgeries and treatments to eliminate her cancerous and precancerous lesions.

I also referred her to an ophthalmologist colleague. Vision problems in albinism result from abnormal development of the retina and abnormal patterns of nerve connections between the eye and the brain, due to lack of pigment. In albinism, the retina, the surface inside the eye that receives light, does not develop normally before birth and in infancy. Thus, the nerve signals sent from the retina to the brain do not follow the usual nerve routes. In addition, the iris, the colored area in the center of the eye, lacks sufficient pigment to screen out stray light coming into the eye. (Light normally enters the eye only through the pupil, the dark opening in the center of the iris, but in albinism light can pass through the iris as well.) It is the presence of these eye problems that defines the diagnosis of albinism. In fact, the main test for albinism is simply an eye exam.

For the most part, treatment of the eye conditions caused by albinism consists of visual rehabilitation. Surgery to correct strabismus may improve the appearance of the eyes, but since surgery cannot correct the misrouting of nerves from the eyes to the brain, it will not provide fine binocular vision. In the case of esotropia, or crossed eyes, surgery may help vision by expanding the visual field, the area that the eyes can see while looking at one point. People with albinism may be either far-sighted or near-sighted, and often have astigmatism.

People with albinism are helped by wearing sunglasses or tinted contact lenses when outdoors. Indoors, a simple change like placing lights for reading or close work over a shoulder rather than in front can make an important difference.

THE COLOR OF CULTURE

Whenever I have patients like Anthony and Frances, it makes me think about skin color and the many factors that determine our destiny. The skin is the most visible aspect of our appearance, and we have a wide variety of genetically determined skin colors. The melanocyte, a tiny structure, is one of the most important parts of the human body for many reasons, including skin color, sun protection, and cultural determination. Many skin diseases can engender heavy loads of social and psychological stress, but those that cause pigmentary alterations are often the most devastating. And even in the absence of such conditions, racial discrimination, wars, and other haunting acts of violence and evil bear witness to how perceptions about skin color plague societies.

All of this for a simple cell. The melanocyte provides us with energy, protection, and the color of our skin and eyes. Melanocytes live in the epidermis, the top layer of the skin, and create melanin, a widely available substance in nature, even seen in plants and fish. The primary role of melanin in the human body is to act as a natural sunblock, absorbing ultraviolet radiation from the sun and quickly converting it into harmless heat. How effective is melanin as a photoprotectant? The photochemical properties of melanin transform harmful UV energy via a process called ultrafast internal conversion, and enable melanin to dissipate more than 99.9 percent of the absorbed UV radiation. Melanin prevents the indirect DNA damage that is responsible for the formation of malignant melanoma and other skin cancers.

Of course, melanin also determines skin color. Imagine you

had a genetic paintbrush, but you only had two colors to mix—pheomelanin (red) and eumelanin (very dark brown). Your choices would be based on the quantity and type of melanin available, which are determined by a handful of genes. One copy of each of these genes is inherited from each of our parents and each gene comes in several different versions known as alleles. Combinations of these genes result in different numbers, sizes, and types of pigment cells, which contain pigment granules called melanosomes and determine the many different shades of skin color. Europeans have fewer, smaller, and lighter melanosomes than people of Africa.

The most recent scientific evidence indicates that all humans originated in Africa and then populated the rest of the world through successive generations. It seems likely that the first modern humans had relatively large numbers of eumelanin-producing melanocytes—and resultant darker skin—as the indigenous people of Africa do today. As some of these original people migrated to areas of Asia and Europe where radiation from the sun was less intense, the selective pressure for eumelanin production decreased.

Some hypotheses about that decrease include protection from ambient temperature, infections, skin cancer, and frostbite; an alteration in food; and sexual selection. Much of the recent research on environmental skin adaptation involves vitamin D production. In northern latitudes, where skin is exposed to meager amounts of sunlight, the inhabitants struggle to make enough vitamin D. Vitamin D is necessary to prevent rickets, a bone disease caused by too little calcium. People with very dark skin require intense sunlight in order to make vitamin D. When people of African descent lived in England during the Industrial Revolution, they were the first to develop the symptoms of rickets, which include slow growth, bowed legs, and fractures. The current theory suggests that at some point northern populations experienced positive selection for lighter skin due to its increased production of vitamin D from sunlight, and, as a result, the genes for darker skin disappeared from these populations. Fortunately, in 1930, vitamin D was isolated and became available as a dietary supplement.

The cultural power of the melanocyte ranges from subtle variations of bias and insult to acts of extreme evil. Diseases such as vitiligo and albinism unfold in infinite variations, depending on the country and the players. We still use color to grade each other as superior or inferior, and make untold decisions based on skin color.

Two of my own most vivid experiences with skin color occurred in Haiti in an orphanage and in Jamaica with skin cancers. During a volunteer medical mission to Haiti in 1980, I and four of my medical school classmates stayed at the St. Vincent's School for Handicapped Children. Besides the school, St. Vincent's also houses a medical clinic that treats about five hundred patients a month, a physical therapy department, an eye clinic, and the only brace shop in all of Haiti. The school is near the capitol in Port-au-Prince. (It was severely damaged during the 2010 earthquake and several staff and children were killed).

I clearly remember walking up to the second floor of the building, to the orphanage and music school, and seeing the children who had been left for adoption. There were cribs, and in each of the cribs were infants—at least five or six in each crib. I remember lifting up several of the children and giving them hugs. I asked about adoptions and the nurse told me, "The light-skinned ones are probably the only ones that will ever be adopted."

In Jamaica, where I go yearly to Treasure Beach, in the parish of St. Elizabeth, to provide volunteer medical services, my colleagues and I treat a group of families that have very light skin and inherently little sun protection. Reportedly, they are the descendants of the survivors of a Scottish ship that was reported to have sunk off the coast of this area in the mid-1600s. After they settled in the area, the inevitable inter-mixing with the local population led to a lineage of residents with light skin, blue and green eyes, and blond and red hair. Others recognize them as "brownins" or "red men" from Treasure Beach. During my last visit I saw five light-skinned members of one family and we removed at least a dozen skin cancers. Can you imagine having light skin and light eyes in the blistering heat of southern Jamaica, and not using any sun protection? The color bias here is one of an enormous lack of information about the viciousness of what the sun can do to those with light skin.

Perhaps one day we will live in a world where color is an objective descriptor and not a defining or limiting character. But for now, we must still struggle against the tragedies brought on by distorted beliefs about skin color and character.

Over a few months, Anthony showed great results with his treatments. He recovered his normal skin color and I continued to follow up on his general health status. Due to her condition, it was difficult for Frances to see the results of her surgeries, but I told her she looked a whole lot better. "That's what people say," she said. "It feels better, too." I checked over the surgical sites. No signs of infection. I reviewed sun protection and care of the surgical sites with her. She seemed pleased, a smile crossing the pink sky of her face.

PART TWO

LIVING IN OUR SKIN

The history of skin and disease is undoubtedly as old as humankind itself. Every culture that was self-aware came to know early on that their skin was host to a plethora of disorders, rashes, and disruptions. A seemingly endless list of discolorations, pustules, plaques, papules, warts, and tumors, each with its characteristic features and signs, can be found in almost any historical medical text from the seventeenth century on.

—Marc Lappe, *The Body's Edge:*
Our Cultural Obsession with Skin

Notions of beauty change depending on social construction. These changes can be evocative, as with skin adornment, or disturbing, as with mutilation. Texture or color or decoration can enhance the features and earn acceptance from others, while other skin manipulations may allow people to be marginalized and stigmatized.

During my over thirty years of clinical work, I have observed how people with abnormalities of the skin cope with daily living. I have treated patients with dozens of protuberant facial neurofibromas, spotted skin from widespread vitiligo, peeling and blistering skin from bullous disease, and socially devastating itch and scaling from psoriasis. In these patients I have observed not only diseases and abnormal anatomy and physiology, but have also noted the racial connotations, metaphorical allusions, and symbolic roles the skin plays as our boundary with the outer world. I am forever amazed at the varieties of afflictions that appear on our skin and how people survive and carry on despite pathologies and social indignations.

The cost and burden of skin diseases are overwhelming. One in every three people in the United States suffers from a skin disease at any given time. Studies commissioned by the American Academy of Dermatology and the Society for Investigative Dermatology showed that skin disease is more prevalent than obesity, hypertension, or cancer, and carries serious medical and financial consequences. They estimated the total annual cost (in 2004 dollars) of just twenty-two skin diseases, ranging from melanoma to acne, at 39.3 billion dollars. The direct medical cost for these diseases totaled 29.1 billion dollars; the total indirect cost associated with lost productivity for these conditions was 10.2 billion dollars. The five most costly categories of skin disease are skin ulcers and wounds, acne, herpes simplex and zoster, cutaneous fungal infections, and contact dermatitis.

Skin diseases can be expensive and time consuming and can affect self-esteem, personal relationships, and careers. Many skin conditions are accompanied by visible physical abnormalities and can impose significant psychological burdens and limit quality of life. Patients with atopic dermatitis, acne, or psoriasis report that their conditions cause greater detrimental impact on their quality of life than do patients with high cholesterol, hypertension, or angina. Skin diseases also have broader health implications, predisposing individuals to infection, scarring, and other diseases.

CHAPTER FIVE

THE BLUE MAN

She had blue skin,
And so did he.
He kept it hid
And so did she.
They searched for blue
Their whole life through,
Then passed right by—
And never knew.
　—Shel Silverstein, "Masks"

I had worked with Patty, the director of nursing at a convales-
cent home where I consulted, for many years. I had learned that
she was not prone to histrionics. So when she asked me to come
by and take a look at Mr. B, I knew his condition would be an
authentic puzzler. What I couldn't have anticipated was just
how profoundly Mr. B would challenge notions of the relation-
ship between our identities and our superficial appearance.

"Dr. Norman, I don't know how to explain this," Patty
began, "except to say that Mr. B has turned blue. He's breathing normal. So I think it's in his skin. Can you come?"

A CALM CHAMELEON

I open this second half of the book with Mr. B not because he was, at our initial encounter, the most anguished of my patients, but precisely because he was very nearly the opposite. In a world of interactions overwhelmingly defined by our visual impressions, we are not supposed to judge books by their covers. Yet we do, inevitably; it is the fate of social animals.

But let's suppose that the object of our judgment were that seemingly least self-conscious of people—one of the elderly? These are men and women who, by and large, are retired not only from the workforce, but also from the dating and mating game. In general, they have accomplished what they are going to accomplish, and they have abandoned, out of some combination of fatigue and wisdom, subtle maneuverings for power and influence. Some have become completely oblivious to them.

For most of us who have not reached that wise age, it is hard to relinquish the superficial and focus on finding serenity for our souls. For Mr. B, for whom self-consciousness was as distant a speck in his rearview mirror as his very youth, it did not seem to be as hard.

I arrived in response to Patty's call during the nursing home's dinner hour. In the dining hall, I found a frail, but alert and courtly, seventy-six-year-old gentleman in khakis and a crew T-shirt waiting for an attendant to bring him his plate of meat loaf, mashed potatoes, and vegetables, and his evening cycle of medications.

I was told that following an earlier stint at the nursing home, Mr. B had briefly rejoined his wife of forty-seven years at their home, before an upper respiratory infection again required full-time medical attention. Prior to that, he had spent his career working for the government, with more than three decades stationed at a post office in upstate New York. His medical history showed, in addition to the respiratory infection that had landed him in the hospital for a week prior to his possible permanent relocation to the nursing home, several episodes of irregular heartbeats. There was also a recent mild stroke, followed by a bedsore on his buttocks that had been infected with MRSA, Methicillin-resistant *Staphylococcus aureus.*

He was on a number of cardiovascular drugs and stroke-prevention drugs. He was also on a long-term course of minocycline, an antibiotic that had been prescribed following the results of the culture and sensitivity tests done from a swab of the bedsore, to wipe out his MRSA infection. I reviewed documentation of the laboratory work and other recent consultations.

Mr B was in no distress. None. Everything about his facial expression and demeanor bespoke, "What's the fuss?"

When I asked him if it was OK to examine him, he became a bit more animated. "You must think I'm some kind of chameleon," he said, and smiled.

"It's a little different than what I normally see," I said. "You are special."

"What's that?"

I repeated myself.

"Oh, yeah." He explained, "I was in the Pacific during the war and worked in artillery. Lost a bit of my hearing."

"I understand," I said, and turned up my conversational volume.

SEARCHING FOR A BLUE KEY

Mr. B had normal vitals. He did show signs of a mild dry skin rash on his legs, and nail fungus. But the most extraordinary thing, what made him special indeed, was that his skin, especially the arms, hands, and legs, was extensively covered with irregular blue-gray pigmented areas significantly altered from his visible pale background skin. His face showed subtle changes, with a slight purple "mustache" and patches under both eyes.

Several suspects needed to be evaluated in the formation of Mr. B's skin discolorations. Any number of problems could cause this kind of abnormal skin coloration.

First, I wondered if his blue nails were by-products of oxygen deprivation. But that didn't square with his blood's oxygen saturation measurement, determined with a finger-probe oximeter. Mr. B's excellent measurement eliminated the ominous possibility of a central or peripheral decrease in oxygen.

Next, I gave some consideration to argyria, a toxic condition resulting from exposure to silver. In the early 1900s, colloidal silver products were sold with silver concentrations as high as 30 percent. Even today in many health food stores and pharmacies, suspensions of silver are sold for the treatment of many disorders, including cancer, acquired immunodeficiency syndrome, sore throats, meningitis, parasites, chronic fatigue, and acne—all without evidence of positive results. It was not inconceivable that Mr. B could have picked up his bluish-gray skin discoloration in this fashion.

But argyria is sun-dependent—in other words, exposed surfaces should have different reactions than nonexposed surfaces. Mr. B's discoloration was uniform.

What about Addison's disease? There are normally two adrenal glands, one located above each kidney. Addison's disease is a severe or total deficiency of the hormones made in the adrenal cortex, caused by a destruction or breakdown of the cortex. Disease in the cortex can affect the hormones that stimulate melanin and can produce a darkening of the skin that may appear as an inappropriate tan on a person who feels very sick. In Addison's sufferers, however, the pigmentation is generally quite regular. Mr. B was not otherwise sick, and his pigmentation was spotty.

Melasma? That is darkish skin discoloration, particularly common in women due to pregnancy or oral contraceptives and generally found on the upper cheeks and forehead. Dirt or extrinsic staining can bring on discoloration and is often confused with a skin problem, and in small areas can usually be eliminated by using an alcohol wipe to remove it. This, too, is generally localized. I crossed melasma and extrinsic staining off the list.

And so it went as my examination proceeded. No vascular rash or any rash associated with porphyria, an enzyme deficiency attacking the skin or the nervous system, could account for the distribution of Mr. B's color. I briefly considered dermatomyositis, a connective-tissue disease affecting muscles and skin, due to the bluish-purple rash on Mr. B's face, neck, shoulders, upper chest, elbows, knees, knuckles, and back. On the other hand, Mr. B had no other symptoms of muscle damage, such as difficulty rising from his chair.

Hemorrhagic rash, brought on by bleeding under the skin that can occur in meningitis and certain hemorrhagic fevers (specifically the type called meningococcal septicemia), was next on my list. A meningococcal rash starts out looking like lots of tiny blood spots under the skin, which increase over time as more underskin bleeding occurs. The spots gradually become bruises, and eventually become large red-purple areas of obvious bleeding. He had none of the signs of meningitis, such as a stiff neck, intense headache, protracted fever, or altered mental state.

Postinflammatory hyperpigmentation can occur following a previous rash or trauma. The resulting discoloration can be persistent, but does not manifest itself with the widespread dusky-blue that Mr. B showed. Nor did his discoloration show any predilection for sun-exposed areas, which is what occurs in photosensitivity reactions.

I next recalled that certain advanced dermatology books describe "blue people" in remote parts of the world, whose color is primarily based on distinct genetic malformations. But Mr. B had spent the majority of his life in Florida and no one else in his extended family had this same problem.

Skin color changes are also associated with hemochromatosis, the most common form of iron overload disease, an inherited disorder that causes the body to absorb and store too much iron. Without treatment, the disease allows extra iron to build up and cause these organs to fail. The skin can show a uniform bronzing; the disease has been given the name "bronze diabetes." But Mr. B's blood results did not show higher than normal levels of iron in the body.

Dissemination of excess melanin through the bloodstream can elicit a widespread discoloration with advanced metastatic melanoma, although it is quite rare. But Mr. B had no melanoma and had irregular, widespread pigmentation changes not seen with this melanoma-induced phenomenon.

Finally, I wanted to make sure to consider whether Mr. B's blueness could be due to extensive ecchymosis, or bleeding into the skin. In smaller, more restricted scenarios, purplish discoloration can represent a small contusion as a result of skin trauma, quite common especially in those who are on steroids that enhance skin fragility. Abuse is not uncommon in the elderly, but the areas of purplish discoloration produced do not generally cover the surface area shown by Mr. B, and there was no history of abuse or trauma identified.

A CULPRIT IN PLAIN SIGHT

To get to the bottom of all this, I asked Mr. B if I could take a small piece of the skin for full laboratory analysis.

"How big a needle you gonna use?" he asked.

"Real small. Just a little pinch."

"Go ahead if it will help figure this out."

I performed a skin biopsy on one of the areas of discoloration and evaluated all my data. After getting my biopsy result back, I knew the answer to this blue riddle. The culprit was not a central or peripheral oxygen thief, a rare blood disease, a hormonal abnormality, the aftermath of a rash, or a residue from a cancer. The culprit was the minocycline medication.

Mr. B had long-term minocycline discoloration—a condition that has been reported in between 4 and 15 percent of those treated with cumulative doses greater than 100 grams. Although a large number of young people take the drug for acne, the elderly are more susceptible to a discoloration reaction.

When I made rounds a few days later, I told Mr. B, "I believe your problem is one of the medications you have been taking.

I'm not sure if your blue color will ever go away, but if we stop the medicine I am fairly certain the color will not continue to spread."

"That's good. It's not a medicine that's keeping my ticker going, is it?"

"No, it's an antibiotic. And if we need to treat you for another infection we have several other ones we can use."

What is it about minocycline that gives it this odd side effect? Genetics seem to play a role, since the abnormality has been reported in twin sisters. A strange phenomenon occurs with minocycline pigmentation—it can discolor internal organs as well as the skin. Multiple cases have been reported of black and green bones, black breast milk, and a black thyroid. The thyroid, in fact, can be knocked into disorder with this phenomenon, and other idiosyncratic reactions can occur such as an illness that resembles serum-sickness, lupus-like syndromes, and autoimmune hepatitis. Gum and teeth discoloration can also be observed, especially with children.

Other medicines, as well, have been shown to increase the risk of discoloration, including the tricyclic antidepressant amitryptilline, neuroleptics such as phenothiazines, and hormone therapy. As I reviewed Mr. B's record, I noted he had not been on any of these other medications, but in fact had been on and off minocycline for many months, for a variety of problems. After I did further wound cultures that showed the MRSA was cleared, I stopped the minocycline. Minocycline is a very effective and widely used medicine, but in Mr. B's case, the discoloration had been an embarrassing though harmless side effect.

I also researched the subtypes of minocycline pigmentation. Type I occurs in areas of scarring, particularly acne scarring. Type II, the type Mr. B had, is a macular outbreak that has a predilection for previously normal skin on legs and arms. Type III produces a brown discoloration with solar accentuation. Each type has its own characteristics based on duration and dosing and reversibility when minocycline is stopped.

When I shared the Mr. B discovery with Patty, the nursing director, she said, "I'm sorry that I was so alarmed. I had only been here a week and nobody had pointed him out to me before. One of the newer nursing assistants told me someone in Mr. B's family was going to pull him out of here and call a lawyer if we didn't find out what was wrong."

Several months later, when I was back in the same nursing home to see another patient, I stopped in to see Mr. B again. He was a pentimento of repair—still fairly blue but faded here and there to regain his normal skin.

"You're coming back around," I said in a loud voice.

"Oh yeah, at least for now," he said. "I wonder what color I'll be next."

CHAPTER SIX

CRY WOLF

Crying wolf is a real danger.
—David Attenborough

"Hello, Mary," I said. "How have you been?"

"Not so good," she said, and handed me a sheet of paper.

Mary looked at me, anxiously awaiting her chance to spill out all her aches, pains, and concerns. This dance between us was a familiar one. I had seen Mary many times over the last five years in my dermatology office. She was a mildly plump, fifty-five-year-old with a daughter who lived nearby. The previous time her chief complaint had been "itchy sores on my scalp." During that visit, she told me she was compulsive, and kept scratching the sores until they bled—a classic sign of the common condition called neurodermatitis, in which the person itches and scratches the skin in an attempt to lessen internal psychic battles. Mary's symptoms were at least in part a result of her psychological condition. When she had come in last a mere touch of the scalp had been enough to make her cry out, and a trip to the salon to get her hair colored had become an excruciatingly painful experience.

She had also started getting infections on her upper back, arms, and face, mostly in the areas where she had scratched herself. "I checked for allergies," she said, telling me that she had gone to the allergist I had recommended to her. "I have an allergy to nickel and to grass—the kind you mow."

But allergies didn't explain her skin problems. Most of them were the lingering effects of her neurodermatitis—her skin had become the target of stress. And Mary had enough perceived and real stress and problems to sink a ship.

I had prescribed a prescription cortisone solution for two weeks and an over-the-counter product that contained castor oil and menthol to cool the fire on her scalp. I also recommended relaxation techniques and counseling, both which often help with neurodermatitis, but our subsequent conversations did not offer much evidence that she had taken my advice.

CARPE ROSTRUM

Now I looked at the list of medications that she had written down—Prozac, Clonazepam, Ranitidine, and others, mostly medications reflecting her treatments for a wide variety of depression and stress-induced disorders. I held up the list like I was looking for suspects in a crime. I was taught from an early stage of my training two things: *primum non nocere* ("first, do no harm") and *carpe rostrum* ("seize the list"). I am always concerned about compliance and drug interactions when a patient has a long list of medications, but I knew from past experience that Mary would go down the list with me, offering a monologue on each drug, and checking if any one of her drugs could be bringing on her current symptoms. Taking control of the list allowed me to focus on her problems and narrow down the suspects.

Included on the paper were small written comments and questions—"Calcium citrate caps—should I take these or not?" and "Need to take this in the morning," referring to her Prozac. Her list of past surgeries ranged from the more trivial ("I had a xanthoma underneath my eyebrow—the doctor removed it right in his office") to a gallbladder removal ("I had two stones"). Over the years she had suffered from a number of problems—tonsillitis, mono, urinary tract infections, blisters, mouth sores and chapped lips, acne, impetigo, high cholesterol, anxiety, and depression. The constellation of symptoms and conditions painted a picture of a stressed and depressed Mary who was chronically taxing her immune system. She seemed to be fighting hard to stay above the water.

Now Mary had a new complaint. "Over the last year I get these bruises all over me," she told me. "They just appear out of nowhere."

Keeping in mind her previous history, I had to consider the strong possibility that her stress had influenced her symptoms. When I asked her if she was experiencing increased anxiety, she was very candid. "I've had severe marriage problems and that has filled my mind with stress and anxiety," she said. "I tried to ignore the problem with the bruising but I keep getting more. The sores on my scalp returned and now I have the bruising. I am very depressed."

I looked at her legs and noticed she had petechiae—red and purple pinpoint-sized dots caused by minor hemorrhages of broken capillaries. The most common reason for petechiae is physical trauma, such as a rough episode of coughing or vomiting, but that usually occurs on the face; these are harmless and generally disappear within a few days. Petechiae can also occur as a side effect of medications such as heparin, sulfa antibiotics, aspirin, ibuprofen, warfarin (Coumadin), and some oral diabetes drugs. Infections such as HIV/AIDS or clotting factor deficiencies can also be responsible for petechiae, as can alcoholism, vasculitis (an inflammation of the blood vessels), certain malignancies such as leukemia or lymphoma, and thrombocytopenia, a blood platelet shortage due to platelet function inhibition.

"Have you had any cuts that have taken take a very long time to clot or stop bleeding?" I asked.

"When I scratch myself, it seems like it takes longer to heal," Mary said.

"Notice any blood in your urine or stools?" I asked.

"Seems like my urine is a little darker," she said.

I queried her about her general health, and in particular symptoms related to her blood functions.

Mary also had small purple bruises on her arms and legs, called purpura, brought on by bleeding under the skin. I ordered some routine tests including a complete blood count, platelets, urinalysis, thyroid, and chemistry profile, and prescribed the medicines that had helped relieve her scalp itch previously.

AN ANSWER IN THE BLOOD

Mary's lab tests came back. Although several of the tests were abnormal, including a small amount of blood in her urine, her platelet count was less than 20,000 per microliter of blood, and that set off an alarm bell.

Platelets, also called thrombocytes, are only about 20 percent of the diameter of red blood cells, the most numerous cell of the blood. The normal platelet count is 150,000–450,000 per microliter of blood, just a tiny fraction of the blood volume. The production of platelets, along with other kinds of blood cells, occurs in your bone marrow. Platelets have one principal function—to prevent bleeding—and stick together, or clot, to seal small cuts or breaks on blood vessel walls.

As with most diseases, the process is one of exclusion. We searched for blood abnormalities other than low platelet count, and any other physical signs beyond the purpura and petechiae. Leukemia, HIV, hepatitis C, lupus erythematosus, cirrhosis, congenital causes, medication problems, and other reasons had to be excluded. Gathering data and funneling down choices over two weeks finally led us to idiopathic thrombocytopenic purpura (ITP), a chronic and noninfectious condition of having a low platelet count (thrombocytopenia) of no known cause. There are approximately sixty thousand adults in the United States who suffer from chronic ITP, with women outnumbering men approximately 2 to 1.

An autoimmune response is believed to be the main culprit in ITP. The immune system normally protects the body from disease and infection. In ITP, the immune system attacks and destroys its own platelets for unknown reasons. As with many of these mystery diseases, there is ongoing research into the cause of ITP.

Two types of ITP exist—acute (temporary or short-term) and chronic (lasting more than six months). Acute ITP, the most common type, generally lasts less than six months, occurs mainly in children, both boys and girls, and generally appears after a viral infection. In children, ITP usually disappears on its own within a few weeks or months and never comes back; the platelet count returns to normal within six to twelve months. A small minority of children, about 5 percent, may need to have further medical or surgical treatment if their ITP doesn't go away.

Chronic ITP may last for years and even people who have severe forms of chronic ITP can live for decades. Although some teenagers and children can get chronic ITP, it mostly affects adults. The severity of the bleeding symptoms and the platelet count determine treatment; in mild cases, no treatment is needed. At some point, those who have chronic ITP are able to stop treatment and keep a safe platelet count. Those with chronic ITP need to avoid medicines that increase risks for bleeding, such as warfarin (a medicine often used to treat an abnormal heart rhythm called atrial fibrillation) and over-the-counter drugs like aspirin and ibuprofen. Alcohol intake needs to be limited because it can decrease the ability of the blood to clot.

ITP can not only result in petechiae and bruising, but can also lead to bleeding from the nostrils and gums and menorrhagia (abnormally heavy and prolonged menstrual period); bleeding into the urine and stool can occur if the platelet count is below 20,000 per microliter. If a person has a very low count (less than 10,000 per microliter, spontaneous formation of hematomas (a localized collection of blood outside the blood vessel) in the mouth or on other mucous membranes and prolonged bleeding from minor lacerations or abrasions often happen. In the rare cases of an extremely low count (less than

5,000 per microliter), very serious and possibly fatal complications can result from subarachnoid (the area between the arachnoid membrane and the pia mater surrounding the brain) or intracerebral (within the brain tissue) hemorrhage, lower gastrointestinal bleeding, or other internal bleeding set off by any minor trauma. Adults are more likely to die *with* ITP than from ITP, due to other conditions that contribute to fatal bleeding.

HEALING FROM THE INSIDE OUT

Over the course of working with Mary, her blood platelets got down to 1,000, a medical emergency. I worked with a hematologist colleague to help get Mary back on track. She was hospitalized several times, and given transfusions of platelets twice. She was also treated with high doses of prednisone to raise the level of her platelet count. The hematologist and I discussed Mary's situation many times, and the drama of the still-lowering platelet count pushed us in the direction of recommending more aggressive intervention—removal of the spleen. During one of the particularly precarious moments in her care, I saw a side of Mary that I had not experienced before—true fright over a very real and potentially devastating disease, as if all of her previous fears were now justified and defined by the fix she was in.

The spleen manufactures the majority of the antibodies that destroy blood platelets; it also eliminates old or damaged blood cells. Since Mary was otherwise medically stable, the removal of the spleen was not an overly serious operation, and she could undergo laparoscopic removal to further reduce surgical risk. However, removing the spleen permanently reduces the body's ability to fight infection and does not fully prevent relapse, although remission following splenectomy is 60 to 65 percent in ITP cases. We talked it over with Mary and her family and she agreed that the surgery was the best way to go.

Once Mary had her spleen removed, she suffered from thrush, a condition in which the fungus *Candida albicans* accumulates on the lining of the mouth. I provided her with antifungal medication and the problem disappeared. She also had a chronic UTI (urinary tract infection) caused by E. coli and was seen by a urologist. Both the thrush and UTI were conditions related to her use of steroids, lowered immune system, and general susceptibility to infection.

"I went through about nine antibiotics before getting rid of the UTI," she told me once she returned to my office, about eight weeks after her spleen was removed. "I still have a problem with urinating a lot, and I stay thirsty all the time." I checked her for diabetes and other problems, but she proved to be relatively OK, and the thirst and frequent urination subsided over time.

After her spleen was removed, her blood platelets went up to 333,000. "So, after that I have had no problems with my blood platelets," Mary said, and her purpura and petechiae cleared up. "I still have the problem with sores on my scalp." She also suffered from low back pain and told me that she had tingling feelings and numbness in her legs, feet, and hands. The surgery had not cured her of all her perceived ills, but it did most likely save her life.

I helped her relieve the pain and discomfort by listening and talking with her about her problems, providing reassurance, and working with her to treat issues as they came up. With every rough journey I go on with a patient like Mary, I hope that it brings us closer and provides the patient with at least a tincture of reassurance that we can help.

While stress could explain some of Mary's skin issues, her petechiae and purpura had been clues to a very real problem. ITP is a serious condition, and although she still had many complaints, due to interventions Mary was alive to tell the tale. In medicine, doctors often encounter these kinds of patients, and their behavior combined with any dismissive attitude on the physician's part can be a barrier to treatment of a serious problem. Being patient with anxious patients like Mary can be difficult, but it is part of doing our best.

CHAPTER SEVEN

STORIES ON THE SKIN

Not one great country can be named, from the polar regions in the north to New Zealand in the south, in which the aborigines do not tattoo themselves.

—Charles Darwin, *The Descent of Man*

One spectacularly sunny Sunday in the end of March, I attended a "tattoofest" in a Tampa hotel. Outside a group of bikers and other attendees sat soaking up the sun. I entered tentatively, not knowing what to expect.

At the check-in tables I picked up a magazine called *Prick,* filled with ads for conventions and tattoo artists' studios. A columnist, Chuck B, wrote, "Even the meanest-looking heavily tattooed characters out there are longing to be coddled, not hurt. Believe me, I know. So let's all get together and have a big love fest."

Inside, I was surprised by what I saw. This was not just displays and sales pitches. People getting tattoos applied and those doing the applications occupied the majority of booths. Music from a blues-rock band, the Accelerators, echoed across the floor, although the din of tattoo machines on high vibration and convention chatter dimmed much of the music.

For the most part, the participants here were quite genial. I took several pictures, after asking permission. A guy named Frank had a tattoo across his upper back: "Love Thy Neighbor." I met a man named Stan who had an ornate design on his chest. "I like dragons," was his summary statement. Sarah had a variety of skin ornaments and a spooky butterfly on her left shoulder.

Howard the Wizard had a wonderful portrait etched into his chest of a man facing a woman holding a heart to her bosom. "That's my wife holding her love for me. She brought color and love into my life."

Gordon had a full body pattern, including a tiger on his head, but there was a nontattooed clear strip up the center of his trunk. "You do not want the center filled in unless you are a gangster leader," he said, referring to the Japanese *yakuza* tradition.

A man named Brian had a flame-embroidered tattoo on his back, bearing the legend "Nephalim," a Hebrew word mentioned in Genesis and Exodus and roughly translated as "giant." He gave me his card, and announced proudly he was doing tattooing in a "major mall."

Kristin showed me a tattoo on her lower abdomen that matched the patterns of her uterus and fallopian tubes and ended in red tulips instead of ovaries. It was truly a striking design, and she turned to show me a smaller, matching flower pattern on her back. "It's a girl thing," she said.

One of the most intriguing people I met was a man named Mike K, who told me he had been in the military for twenty-three years and had retired four years ago. He had never had a tattoo until two years ago, and his first was an eagle on his right shoulder. The bird was now hidden in a tapestry of colors that covered his entire body. "Do you have any particular theme?" I asked. "Yeah," he said. "After the eagle I got into evil, the faces of death." He pointed to a black widow spider on his chest. "I like the dark side."

The electric tattoo machine was patented in 1891, but tattoos have existed for thousands of years. Traditional tattooing methods have included sewing lines into the skin (Inuit), puncturing the skin with a long sharp point (Cambodia, Burma, Thailand, Native America, and ancient Europe), raking the skin with rows of needles (Indochina), chiseling the skin by hitting the chisel with a hammer blow to create precise lines (Samoa), and hand-poking it with a *tatu* needle (Japan).

The modern electric method certainly causes less trauma, although it can be quite uncomfortable. I observed the modern process, in which a small machine with one or more needles connected to tubes containing dyes makes multiple injections into the dermis. With each sweep of the needle, the tattooist guides the machine over the skin and controls its speed. A session can take several hours, depending on the size of the tattoo.

Most of the tattooed people I met also had various piercings, mostly (so far as I could see) in the lips, ears, nose, and nipples. Included in the festivities, however, was a half-hour-long piercing demonstration in which a purple-haired artist, providing no anesthesia, slipped eighteen-gauge needles attached to feathers into the back skin of a young fire-haired woman. She subsequently became a human peacock, with her new appendages shooting upward and outward in a proud display. I asked the artist if he was patterning this after anything in particular, and he said, "I've got a vision in my head."

In *Skin: A Natural History,* Nina Jablonski writes that "human skin is unique because of what people do with it. Among other animals, the skin and its appendages, including scales, feathers, and fur, serve to advertise an animal's anatomy or prowess. But the human penchant for decorating the skin is unmatched in the animal world. A peacock can flaunt its feathers, but it can't change them for every show. The colors or pictures we apply to our skin communicate our values and aspirations as well as our hopes and personal histories." Jablonski notes that even when we adopt the "natural look" and don't adorn our skin at all, we are making a social statement: "Our skin talks even when we don't; it is not a neutral canvas. Through the expressive functions of skin and body decoration, we have expanded the communicative potential of our bodies and reinforced the primacy of the visual sense in our sensory repertoire."

I did not stick around for the various awards given to the best tattoos and artists. As I drove away, I had mixed feelings. Part of me wanted to take a shower and the other was fascinated by the display of skin and art. I knew that I would never look at a tattoo in the same way again; this love fest had forever etched into my mind the extraordinary methods and patterns of the art of tattooing.

CHAPTER EIGHT

DOGGIE NIPPLES AND EXTRA EARS

Let us a little permit Nature to take her own way; she better understands her own affairs than we.

—Michel de Montaigne, "Of Experience"

Christie, a twenty-one-year-old, came to my office because, she said, "I have moles on my body that bug me." She showed me a few on her neck and back, then shifted my attention to her abdomen.

Below each of her breasts, in line with her areolas, were small dark nipples, each a bit less prominent than the ones that topped her breasts. "Oh yeah, my doggie nipples. I think they said my mom took some pill for morning sickness that may have affected me. It's not that they bother me that much except they grew during my pregnancy." She was concerned about one particular mole on her neck and so was I, so I biopsied it to check for abnormality. It turned out to be a bit irregular but not life threatening.

·   ·   ·

"I think my son is growing another ear," the mother said when she and her son, Brandon, arrived in my office. Indeed, her son did have an accessory tragus, a bump of slightly oblong skin that rose about five millimeters above the surface in front of his left ear and had the same tone as his flesh. The child and his mother both wanted it off. He was at that age when his schoolmates would point out and descend upon any incongruity in bodily shape or form with ruthless indiscretion. Part of the delay in getting it removed was due to his medical insurance authorization but another part was the mother's request: "My mother is coming from Puerto Rico and she said I had the same thing when I was small. So she really wants to see it before you take it off."

I took if off a month later, after Brandon's grandmother had a chance to inspect the family birthmark. The result was a child happy to not have a "third ear."

MORE THAN ENOUGH: REDUNDANCY AND THE SKIN

I have always been fascinated by redundancies in nature such as extra nipples and ears. Over the years I have seen duplicate ureters and kidneys, extra digits in humans (and in cats), and duplicate blood vessels. I have noted that such duplications engender tales, as when the position of a redundant organ fools a surgeon in search of standard landmarks.

Supernumerary nipples (SNNs) are the most common form of accessory mammary tissue. The growth results from the incomplete regression of a portion of the milk lines, the two lines that are developed along the frontal surface of the embryos of mammals of both sexes, from the arms to the legs, and give rise to the mammary glands and nipples. Although SNNs usually occur sporadically, up to 6 percent of people with this condition have a family member who also has it, and inheritance in an autosomal dominant pattern has been suggested.

Depending on sex, ethnic group, and region of the world, there is a wide range of prevalence of SNNs, from 0.2 percent to 5.6 percent. Some studies suggest that men are affected more than women, but other studies suggest each sex is equally affected. For unknown reasons, black infants are affected up to eight times more than white infants. Up to one-third of people with this condition have two SNNs, and 2 percent to 4 percent have three or more SNNs. They are most common on the left side of the chest—especially below the nipple—and are seen in various sizes. Supernumerary nipples can occur in unusual places such as right next to the primary nipple, over the shoulder blade, on the neck, or as a mass on the vulva.

In 1915, Kajava classified SNNs into eight groups, and these classifications are still used today. The types range from the most common, the redundant nipple only, which is known as polythelia, to a patch of hair only, which is known as polythelia pilosa. A redundant areola only is known as polythelia areolar, and a redundant nipple, areola, and glandular breast tissue is known as polymastia.

SNNs are usually asymptomatic and can be left undiagnosed for years. In rare cases, SNNs can undergo pathologic changes; there are reports of Paget's disease of the breast (a cancer) and lipoma, a benign fatty tumor, being associated with SNNs. Researchers have also suggested that SNNs may be used as a cutaneous sign of underlying abnormalities, although that pattern is far from absolute. Black infants with SNNs, for example, have not been shown to have anomalies associated with their SNNs.

Lactating women with SNNs may secrete milk from their accessory nipples. When SNNs have very little ductal or glandular tissue, lactation will come to an early halt; if SNNs are more developed, they may secrete milk continuously during lactogenesis.

Of course, there may also be psychological costs. I didn't ask Christie this, but I've read that certain women with SNNs internalize a sense of either hyper-femininity or freakishness. How do people cope with having bodies that aren't "normal" in our culture? What costs does this exact?

During one of my medical mission trips to Jamaica I took care of Paula, a thirty-two-year-old woman who complained of having an "extra part" on each of her hands. Upon examination, I noticed a redundant finger attempting to grow under each of her little pinkie fingers, along the side of her hand. We discussed the possible issues of their continued growth and she was more than anxious to get them removed. Although she appeared a bit scared, I took off the first one in a short time. Before taking off the second one, I raised my hand and said, "Give me a high six!" and our hands joined. The medical student assisting me laughed and Paula smiled, too. It seemed to calm her down, and within moments the second appendage was off and in the biopsy bottle. In one of my next visits to Jamaica I removed an extra digit on a five-year-old from Paula's family.

You can call the condition that Paula and her relative experience polydactyly, polydactylism, or hyperdactyly, but in any case, it translates to the anatomical abnormality of having more than the usual number of digits on the hands or feet. People with polydactyly have six or more digits on their hands, their feet, or both. Each extra appendage varies, from small pieces of soft tissue to apparently complete digits. Most are surgically removed early in life.

The condition is reported to occur in about two children in every thousand but the frequency varies greatly. In the United States, the Pennsylvania Dutch have a markedly increased incidence of polydactyly. This rate is due to the founder effect, a phenomenon that occurs within microevolution, when an isolated environment is occupied by only a few members of a species who then multiply rapidly, resulting in a sharp loss of genetic variation. The new and genetically protected population may be distinctively different from its derivative parent population, both genetically and phenotypically. The raised probability of inbreeding results in an unusually high number of genetic defects, including polydactyly.

The historical records on the phenomenon of polydactyly are quite varied and filled with stories of genetic wonder, particularly in places where consanguineous marriages were prevalent. Within the Hyabite tribe in Iraq, in the family of Foldi, the members were reported to limit their marriages to within their tribe. Each child was reported to have twenty-four digits, and infants born with the normal number of twenty were sacrificed as being the offspring of adultery, according to early visitors that observed the tribe. Voight recorded encountering a patient who had thirteen fingers on each hand and twelve toes on each foot. Saviard reported having seen an infant at the Hôtel-Dieu, the ancient public hospital in Paris, in 1687 who had forty digits total, ten on each hand and foot. Annandale related the history of one woman with six fingers and two thumbs on each hand, and another who had eight toes on one foot. Meckel told of a case in which a man had twelve fingers and twelve toes, all well formed, and whose children and grandchildren all inherited the deformity.

Of course, it's not only humans who experience polydactyly. During off hours at a medical meeting in Key West, I visited Hemingway's old digs to explore and to see where he did his writing. And of course, there were the famous cats to check out. Ernest Hemingway reportedly was given a six-toed cat by a ship's captain and developed a fondness for this cat and others with this trait. When Hemingway died in 1961, his former home in Key West became a museum and a home for his cats. Today there are more than fifty descendants of his cats, and half of them have polydactyly. "Hemingway cat," or simply "Hemingway," is a slang term that has come to describe feline polydactyly. Although normal cats have a total of eighteen toes, with five toes on each front paw and four toes on each hind paw, polydactyl cats may have as many as seven digits on their front or hind paws.

THE EVOLUTION OF A PERFECT TEN

Where humans are concerned, I certainly could imagine some practical advantages to having extra digits—fourteen fingers so I could type this book faster, an extra arm for a quick overhead shot in tennis, or an actual third eye on the back of my head for dealing with busy traffic. In the sweep of evolutionary time, one might expect some species to have found a survival advantage in having extra digits—perhaps in improved tree climbing or toolmaking.

But when we compare species, we find that though many land animals today have fewer than five digits per limb, there is a notable absence of species with more than five. From an evolutionary perspective, losing an unneeded feature is relatively easy compared to gaining a potentially useful feature. What's more, extra digits might not be all that useful. Clifford J. Tabin, a professor in the department of genetics at Harvard Medical School, has suggested that the genes that control digit formation are structured to produce five basic patterns, in the case of a human hand—thumb, pinky, and middle three fingers. Additional digits that arise through genetic abnormalities simply duplicate an existing digit and offer little species improvement or survival benefit.

The most daunting genetic constraint on polydactyly may arise from pleiotropy, the capacity of genes to influence multiple physical characteristics. A rare genetic disease called Hand-Foot-Genital Syndrome appears to arise from incorrect coding across the genes that handle digit formation, and involves malformations of the genitourinary system as well as limbs. Given this, it may be that during the course of evolution, any advantage that came from having more digits also included the disastrous disadvantage of having malformed reproductive organs. Survival of a person with extra digits could have limited species survival.

Our bodies certainly do contain vestigial organs, organs and structures that no longer appear necessary but have not entirely disappeared. Vestigial does not necessarily mean without function; it is quite difficult to ascertain with certainty that any particular structure is actually functionless. Perhaps its use simply is not recognized yet or it is still functioning but greatly diminished. If a human organ, structure, or behavior is to be labeled vestigial, it needs to be homologous with a functioning organ in other species.

For example, all tetrapods—animals with four limbs—have pelvic bones. Pelvic bones are needed to help support lower limbs, yet whales still have pelvic bones, as do snakes and legless lizards. Why do they have pelvic bones if they don't need pelvic bones to move? The sensible explanation is that all tetrapods evolved from a common ancestor.

To give another example, the appendix in humans has long been considered a vestigial organ, but recent evidence states that it may perform certain immune functions. When I was fifteen years old, my appendix burst and I was rushed to the hospital for emergency surgery: certainly this would appear to be a disadvantage for species survival. Again, common sense points to evolution and a common ancestor.

In 1893, the German anatomist Robert Wiedersheim listed eighty-six alleged vestigial organs in humans, including wisdom teeth, nipples on men, and the appendix. He was following up on the list of "rudimentary" organs that Darwin described in *The Descent of Man*, which included the muscles of the ear, the tail bone, and body hair. Wiedersham used his list as an argument for evolution and described the organs as evolutionary leftovers with no current function. However, the list contains structures that today are known to be essential; it is a historical record of that day's physiologic knowledge.

.   .   .

As for Christie, she never mentioned any further concerns about her extra nipples. Nor did I ever see Brandon and his mother again. But when I return again to Jamaica, I may encounter another polydactyl member of Paula's family who will want an appendage removed. Now that I have researched all this, I will have an extra tale about extra body parts to tell while I am performing the surgery.

CHAPTER NINE

AT WAR WITH OUR SKIN

All the wise world is little else, in Nature,
But parasites and sub-parasites.

—Ben Jonson, *Volpone*

My patient, Brunelle, a fifty-four-year-old, told me her story.

"I'm real stressed. I'm so stressed I get an irregular heartbeat. I've been to four doctors already for this problem and none of them said they can help me."

Before I could respond, she leaned forward and showed me her scalp.

"I've had an infection in my scalp," she said, moving toward my face. "That's been going on more than two years and I've lost lots of my hair." I examined her face and noted that large areas on her face appeared to be scratched. "I got bugs on my scalp and so I shave my head to try to get rid of them. She yanked hairs off her scalp and put them on a white sheet of paper to show me. "These were like crabs. I could hear them crack when I mashed them with my hands."

ENEMIES WITHIN

Approximately 30 percent of all dermatological conditions are associated with a psychiatric disorder. Recently I started using a psychological inventory, a revised depression assessment scale, to evaluate these patients. Brunelle filled out this scale, with its fifty questions, each with a response to be circled from one to four. Four indicates intensely bothersome symptoms and one the least problematic. I looked at the results when she finished. She was highly positive (she wrote "4+") for muscle pain and joint pain, as well as difficulty breathing and constantly feeling stressed. In almost every category, she marked herself a three or four.

"When I was first starting to get these, I used a Dustbuster and vacuumed my skin. I was wearing a bandana to keep them from falling into my eye."

Brunelle used a black-tipped cane and wore a patch over her left eye. I asked her about her eye, and her husband, a mid-sized mustachioed man with a rough complexion and tangled hair, offered his report. "She fell on the edge of the counter. Blunt force trauma. Doctor told us it severed the retina."

"Awful," I said.

"Doc, I gotta ask you," he said. "Is it contagious? I don't have any problem."

"I don't think what she has is contagious," I said.

"We live way out in the woods and I thought maybe it was from the bugs out there," he said.

"Do you remember when it first started?" I asked.

"Sure do," she said. "I got a plant from Publisher's Clearing House. That was 'bout three months ago. It had a fungus on it that smelled like dead deer with maggots. I think that is what did it. I looked at the plant and it had oblong, liver-colored blotches on it."

"How big do these critters on your skin get?" I asked.

"Doc," she said, and pinched her thumb and index finger to give me an approximation. "Maybe an eighth of an inch long. They started up and down my leg and they leave red circles. Then a slit opens up and they come out. I put my soul on the line that some look so big I could go fishing with them. I go to get them and they sink down into my skin. Way down under."

"Do they bother you more at any particular time of day or night?"

"They are worse at night. Never itch but I still feel them. I put a Band-Aid on them because they won't stop producing," she said. "I had the exterminators at the house twice and they sprayed everywhere. I still got the bugs all over me."

Over Brunelle's next two visits I took multiple skin samples and discovered no evidence of any arthropods or critters on her scalp. In addition to the skin scrapings, we ordered a blood count, thyroid function measure, vitamin B-12 levels, and a few other tests. I asked her about getting psychological therapy, and she refused. I prescribed her a topical steroid cream and an over-the-counter product with menthol and camphor that I knew would help soothe her skin. Once I had reviewed her medical history and found her to be a candidate, I also prescribed a drug called pimozide to calm both her nerves and her itch.

"This will help you get rid of the bugs," I told her. I tried again to get her to counseling but she never went.

I saw her again four weeks later, and we made a plan for me to work with her primary care doctor to coordinate her use of these medications. On her next visit, six weeks later, Brunelle filled out another inventory. The results were refreshing; she had shifted her responses to the lower end of the scale, the realm of a more bearable life. "You saved my life," she said. "No one else would listen to me. All the others said I was crazy."

NEURODERMATITIS: DELUSIONS, OBSESSIONS, AND UNSTOPPABLE ITCHES

Brunelle had delusions of parasitosis, a condition that can be misdiagnosed or overlooked, often for years, though the disease was first described in scientific literature in the 1800s. I have had hundreds of patients who have been treated with topical medicines for this condition by physicians who assumed they had scabies or another malady. These patients were often prescribed, perhaps honestly, dozens of concoctions that never resolved the underlying problems. Physicians who treat these delusional patients can feel frustrated and angry because the patients are so insistent about having infestations and the physicians can't find any. The patient may feel more isolated and confused with each visit until an accurate diagnosis is established.

As humans, we always seek an explanation for our problems. According to the Centers for Disease Control and Prevention (CDC) website, under the category Unexplained Dermopathy, it states that "the CDC continues to receive reports of an unexplained skin condition which some refer to as 'Morgellons.' Persons who suffer from this condition report a range of cutaneous or skin symptoms including crawling, biting and stinging sensations; granules, threads, fibers, or black speck-like materials on or beneath the skin; and/or skin rashes or sores. In addition to skin symptoms, some sufferers also report fatigue, mental confusion, short term memory loss, joint pain, and changes in vision." Brunelle hadn't called hers Morgellons, but this is otherwise a fair description of her symptoms.

Over the course of my more than twenty-five years in the skin trade, I have encountered all kinds of weird pathologies, a small fraction of which are in the category of severe neuro-dermatitis, a psychological process with dermatologic mani-festation. I once had a patient in a nursing home named Ricky who had unexplained weight gain. Upon inspection, the nurs-ing assistants noted that Ricky continually ate his T-shirt. The nursing director decided he was getting too much starch in his diet via his bizarre habit and that is what brought on the excess weight. He also scratched off parts of his scalp and ate them and had to be physically restrained to prevent him from dig-ging into his neck and cutting into his jugular vein.

The markings from neurotic scratching can result in div-ots and skin atrophies that imitate other skin diseases. And the reverse is true: I have seen many patients who appeared to have self-induced scars from "nervous picking," but who I discovered, with further exploration and a biopsy, had cutane-ous lupus or other problems. As with any suspected disease, it is important not to tunnel the diagnostic vision too rapidly. In each situation, a thorough history, physical examination, and diagnostic tests may be needed.

Factitial dermatitis or dermatitis artefacta, such as acne excoriée, occurs most often in patients in their early teens and early adulthood. This psychocutaneous disorder is produced by the person's own actions—a deliberate and conscious production of self-inflicted skin lesions—and often represents a maladaptive response to a chronic or acute psychosocial stressor such as emotional deprivation, underlying anxiety or depression, an unbalanced body image, or a borderline personality disorder. These skin lesions serve as a nonverbal and dynamic cry for help. Chronic medical or dermatological conditions are seen in many dermatitis artefacta patients. The disease is poorly recognized and very underreported. With repeated episodes of self-mutilation, the person acquires disfiguring scars on exposed areas of the body.

It is imperative that medical practitioners understand and recognize neurodermatitis, psychocutaneous diseases, and other skin conditions that are psychologically related, especially given their prevalence and the destruction that they cause. The skin conditions whose underlying cause may be the result of a psychological condition include delusions of parasitosis, dermatitis artefacta, lichen simplex chronicus, neurotic excoriations, and prurigo nodularis. Myriads of other diseases, such as psoriasis and eczema, are ignited by stress and other psychologically related fires.

Among the latter category of pathology are skin conditions that reflect underlying substance abuse. I remember a twenty-eight-year-old man named Jeff who was referred to me with a three-month history of recurrent blistering and nonhealing wounds on the right knee. He complained of numbness on the affected area but repeated objective examination did not reveal any sensory deficit. Routine laboratory tests were normal and a skin biopsy only showed evidence of scratching.

I treated him with topical steroids and antibiotics and an oral anti-itch medicine. He noted new lesions on other areas. During evaluation he told me, "My wife and I are not getting along very well." I recommended psychotherapy, antidepressants, and partner counseling. "I scratch all the time," he said. "I put on the cream every thirty minutes. I set the alarm on my watch." As he sat in front of me he showed fidgety and obsessive behavior and would ramble on and on about rather mundane problems in his life.

"I do pick at my skin until it bleeds," he said. "I got bugs crawling under my skin and I got to get them out."

Within the first few visits I suspected that Jeff was a methamphetamine addict. Along with the more obvious signs such as "meth mouth" (dramatic tooth decay and halitosis due to horrible hygiene), he had skin lesions that took a long time to heal. Meth inhibits the body's ability to fight off minor infection, so Jeff was always bringing on more distress by scratching his skin. Delusions of parasitosis are another symptom of meth addiction (and also of cocaine addiction), and Jeff had this turned on full volume. As with most addicts, the intellectual violence of the ego can turn into a physical assault on the skin.

What brings on this war against one's own skin? I have had hundreds of patients with skin hypochondriasis, who, like Jeff, think their skin is "diseased." As with many diseases, there are multifactorial causes that include genetics, psychosocial factors like addiction, and a personal or family history of psychiatric illness. Fried and others created categories for the different types of skin pickers, including "angry, anxious/depressed, body dysmorphic, borderline, delusional, guilty, habit, narcissistic, obsessive compulsive, and organic."

. . .

I remember another of my patients, a sixty-year-old female whom I will call Mrs. A, who told me, "I feel as though I am infested with bugs that are constantly biting me." She claimed that the bugs had been all over her house for many weeks and that the infestation was so bad she might have to sell her house. I noticed scratch marks on her arms and face but did not see any signs of bites, burrowing, or infestation. Upon further questioning, I found that her medical history was unremarkable and she seemed to be in otherwise great health. She was on no medications, had no known allergies, and denied any use of drugs. Mrs. A said that she kept her house as clean as possible and took at least two showers a day since the infestation. She had not been outside except to go from her car into her house or work. "When I am at home, the bugs follow me wherever I go. If I move, I will have to find a way to trick the bugs. The bugs are taking over my life."

Such delusions can take a powerful hold. Regular psychotherapy might help Mrs. A deal with the life event that triggered the delusion, but psychotherapy alone very likely would not eradicate her profound belief that a parasite is attacking her. New research shows that these delusions arise from a chemical imbalance in the brain, which explains the power of pimozide.

Pimozide is an antipsychotic medication that works by altering brain chemistry, specifically by affecting the neurotransmitter dopamine. Pimozide is also used to suppress the motor and phonic tics associated with Tourette's syndrome. Because the drug is a dopamine antagonist, and can produce side effects including various movement disorders such as stiffness and restlessness, I usually start treatment, as I did with Brunelle, at one milligram daily and check back in a week. If needed, I will increase the prescription by one milligram increments every week until either optimal clinical response is reached or patients are taking four to six milligrams per day.

Many drugs and other therapies can be effective for bouts of neurodermatitis. Atarax and doxepin are anti-itch (antipruritic) drugs that may be helpful. When neurodermatitis is linked with obsessive-compulsive disorder it can be treated with selective serotonin reuptake inhibitors such as fluoxetine or sertraline. The treatment method includes instructions about not scratching and symptomatic treatment with topical steroids and moisturizers. Among other frequently prescribed medications for the treatment of psychosis are newer atypical antipsychotic agents that are dopamine and serotonin receptor antagonists, including risperidone, olanzapine, and quetiapine.

Nonpharmacologic treatments include bandages, biofeedback therapy, relaxation training, hypnosis, and psychotherapy. All therapies need to be done in an accepting, empathetic, and nonjudgmental environment with mutual trust and rapport between patient and doctor.

Some forms of neurodermatitis are driven by obsession or compulsion, two different manifestations of the mind-body, brain-skin interaction. Compulsion is a repetitious behavior that proves hard for certain people to suppress, and obsession is a repugnant thought that intrudes repetitively into a person's mind. If either the obsession or the compulsion is of sufficient intensity to interfere with a patient's lifestyle or cause significant subjective distress, the diagnosis of obsessive-compulsive tendency may be justified. A person may have either or both. Some people may obsess about bugs crawling under their skin or bad acne, but never touch or scratch in a repetitive way. Others may have an irresistible compulsion to pick their skin, but do not describe any recurrent thought process associated with their compulsion.

Within dermatology, obsessive-compulsive tendencies can be noted in many circumstances. Hair loss may be brought on by trichotillomania, compulsive hair pulling. Excessive washing of the hands can elicit hand dermatitis. Scratches and subsequent chronic skin infections can be related to neurodermatitis and its attendant compulsive skin picking. Without appropriate treatment, obsessive-compulsive behaviors generally have a chronic and often debilitating course.

There are important differences between these obsessions and compulsions, on the one hand, and delusions on the other. Both involve mental preoccupation with an overvalued idea. But most adults with obsessive-compulsive tendencies recognize and acknowledge the unreasonable, excessive, and destructive nature of their obsessions and compulsions, though they are unable to stop. I have had many patients with acne excoriée or other forms of neurodermatitis who tell me, "I know I should not be picking my skin, and if I keep doing this I'll get scars, but whenever I try to stop, something drives me to start again."

By contrast, people suffering from delusions have faith in the validity of their delusional thoughts. Most are so convinced of the reality of their ailment that they first seek help from dermatologists, not psychiatrists. Patients have brought me in skin scrapings in matchboxes, plastic bags, and even on microscope slides to demonstrate the reality of the imagined insects. Although the obsessive-compulsive tendency and delusion are often easily delineated, at times this difference about belief can be a challenge to differentiate.

As Sherwin Nuland writes in his book *The Soul of Medicine,* "Like the mind, the nervous system is everywhere. Any little manifestation of its possible tics seems to be transmitted to each small area of the thinking and doing parts of us that allow little rest from its manifestations. And those thinking and doing parts respond with messages of their own, so that neurological and mental diseases influence every fiber of what we are, too often in ways about which little can be done. So the person adapts, and it is in the adaptation that so much of the difference is seen between one person and another."

HEALING THE EXTERNAL NERVOUS SYSTEM

Why is the skin so connected to the nervous system? The skin, in common with the nervous system, arises from the outermost of the three embryonic cell layers, the ectoderm, which is the general surface covering of the embryonic body. The central nervous system's primary function is keeping the organism informed of what is going on outside it. It develops as the inturned portion of the general surface of the embryonic body. The rest of the surface covering, after the differentiation of the brain, spinal cord, and all the other parts of the central nervous system, becomes the skin and its derivatives—hair, nails, and teeth. The ectoderm also gives rise to the sense organs of smell, taste, hearing, vision, and touch—everything involved with what goes on outside. Thus, one might consider the nervous system a buried part of the skin, and the skin might be regarded as the external nervous system. These two organ systems begin in intimate association and remain interconnected throughout life. As Frederic Wood Jones, the early twentieth-century English anatomist, put it, "He is the wise physician and philosopher who realizes that in regarding the external appearance of his fellow men he is studying the external nervous system and not merely the skin and its appendages."

I recently attended a lecture by Miguel Nicolelis, a co-director of the Center for Neuroengineering at Duke University. He had just written a book, *Beyond Boundaries,* in which he explains that "the long-sought merger of brains with machines is about to become a paradigm-shifting reality." I saw some of that reality in the video portions of his lecture, in which research monkeys moved objects on a computer screen using neuroprosthetics. As the copy for his book stated, "Imagine living in a world where people use their computers, drive their cars, and communicate with one another simply by thinking." Nicolelis's work details the boundaries of our humanness and how we can combine technological advances to create humans with new and perhaps incredible changes in their sensory and motor input and longevity. I find it fascinating to consider the physical and psychic boundaries of our own skin.

I have always been a big admirer of those in the research realm of neurology, who study our amazing brain. While listening to Nicolelis's lecture, I thought about what our mind contains and how we can quantify our activities and beliefs. In reflecting about what science does and does not know about brain chemistry and neurodermatitis, I catalogued a number of unanswered items, including what happens inside the mind of a person with delusions of parasitosis. Someday we will perhaps know what part of our nervous system engages and how it connects with other components in us when we do a moral or courageous or obsessive or self-destructive act, and the minimal neural ensemble size required to create a belief or a behavior. As Nicolelis describes it, the brain is the only organ that both plays music, through its vast array of electrical signals, and reorchestrates the instruments that play the music. The brain is not impassive; it always has a point of view and we are not certain which cells are responsible for it. We still don't know what kind of music plays in the brain of someone suffering from a debilitating neurodermatitis, and how pimozide and other drugs help rearrange the music of thoughts and provide relief.

As Marc Lappe writes, "The skin itself is also a common target of deep-seated psychoses and neuroses. This is so not simply because it is 'reachable,' but because of the deep roots that connect a sense of self to the sensorium. For some of us, the skin becomes a focus for obsessive preoccupation."

The psychiatrist attempts to read the mind, listen, and treat. The neurodermatologist must read the skin and the mind and be a master gardener of both. When it comes to the concept of delusions of parasitosis and other neurologically influenced skin diseases, perhaps the most important question is this— how far can you take your sense of self? Lappe continues, "Many individuals struggle with a literal inability to 'sense' the self. With the exception of vague proprioceptive sensations that are generated by the intestine and some viscera, the only place where the mind can sense the body is through the skin. This reality often makes the skin the primary canvas on which our cultural and personal identity is drawn."

. . .

I have always liked the accessibility of the skin and how a careful reader can interpret the skin's vagaries. Observation of the skin of someone like Brunelle, Jeff, or Mrs. A allows us to detect signs of underlying systemic and psychiatric disease, to help those who suffer gain an understanding of their puzzling symptoms and, hopefully, to get better.

CONCLUSION

THE FUTURE OF OUR SKIN

As the most self-involved and manipulative of primates,
we will continue to change ourselves to an even greater
extent than we change the world around us. We cannot
predict our future skin in detail, but major trends in
medicine and art provide us with some leads.

Nina Jablonski, *Skin: A Natural History*

In *Skin: A Natural History,* Jablonski suggests that the functions
and potential of skin will be expanded along at least three
major frontiers in the next several decades, by a wide variety
of constituencies.

She describes the first frontier as an alteration of the bio-
logical functions of skin in order to treat specific diseases and
injuries, largely driven by the work of specialists in gene therapy, molecular biology, pharmacology, and bioengineering.
The second major frontier will, she says, focus on communication through and with innovations, such as implanted sensors
and communication devices. In this realm, aesthetic innovations will generate new types of cosmetics and skin colorings.

"When it comes to our skin—our protective envelope, billboard, and largest sensory portal," Jablonski writes, "our scientific and artistic creativity knows few bounds."

In fact, according to Jablonski, new combinations of practical and aesthetic considerations are likely to yield the most surprising changes in human skin–based communication, including projecting visual imagery onto the skin or implanting sensors capable of communication and stimulation. The interaction of implanted devices with visual and auditory devices, including interactive virtual reality, may lead to unique and highly entertaining experiences.

Jablonski proposes that the third major frontier for future skin will be the creation of "skin" for robots, to allow them to simulate human touch. Using that skin, computer engineers and psychologists will be able to create interactive virtual reality experiences.

These and other innovations will extend our current thinking about what skin is and what it is capable of and will, Jablonski says, "challenge our very notion of corporeality." She notes that our future skin will continue to do what skin has been doing quite proficiently for us, but with added technology it may "extend our touch and consciousness far beyond our own skin."

Dermatological concerns—hair, skin, and nails—are among the most common reasons for consulting a physician and pharmacist. Yet we certainly need to make changes in how we treat dermatological afflictions. Many of the most common skin conditions have been diagnosed since antiquity and many as yet have no cure. Fungal infections, psoriasis, eczema, skin cancers, and warts have stymied physicians and researchers for generations.

Therapy in dermatology, particularly in the treatment of psoriasis and eczema, is changing significantly as new approaches to helping patients access treatments and already-marketed products find new uses. As a result of the increased understanding of the molecular mechanisms of skin diseases, dozens of drugs and laser therapies are in phase II or III trials. The "survivors" in this arduous contest will reach the market in the near future.

Though psoriasis, with its attendant chronic itchy, silvery-white, scaly plaques on elbows, knees, scalp, and elsewhere, has plagued humans for millennia, even recent therapies often require lengthy treatments to provide relief that is often only temporary. Patients may be treated with some combination of topical steroid, vitamin D-3 analog, anthralin, light therapy, tape or occlusive dressing, and intralesional steroids. Patients with more serious cases and a higher percentage of affected body surface can also be treated with oral therapies such as etretinate and oral steroids. Though these methods may achieve some success at reducing the plaques and itchiness, relapses are expected, no cure is accomplished, and the possibly severe side effects of the drugs are quite worrisome.

Methotrexate, another conventional psoriasis therapy first introduced as a cancer therapy, is an anti-metabolite that has shown some success in the treatment of pustular psoriasis and psoriatic arthritis. Psoriatic arthritis can occur in up to 30 percent of those with psoriasis and includes joint and connective tissue inflammation and the pain of arthritis along with the skin disease. Methotrexate acts by reducing inflammation, histamine release, and skin cell proliferation. However, patients taking methotrexate must be carefully monitored for multiple systemic side effects.

A new wave of psoriasis therapy began with the observation that cyclosporine was effective for treating the disease. Cyclosporine, a powerful immunosuppressant traditionally used to prevent rejection of transplanted organs, is used to treat plaque psoriasis, the most common form of psoriasis, affecting 80–90 percent of people with psoriasis; the victims may suffer with widespread inflamed skin and itchy plaques. Patients with psoriatic arthritis and psoriasis of the nails can also benefit from the drug, though they often require longer treatment periods. The success of cyclosporine led to the development of medicines that target the immunologic and inflammatory cascades, including agents called "biologics" that interrupt the T-cell interactions that contribute to psoriasis formation. Biologics have started to dominate the psoriasis market share since their introduction in 2003, skyrocketing to at least half of the market today.

New treatment options are also on the way for xeroderma pigmentosum (XP), whose young bearers have severe solar damage, skin cancers, pigmented dry skin, and eye abnormalities. Xeroderma pigmentosum is an autosomal recessive disease, characterized by defective DNA repair. New therapies for this disease and many other skin diseases are based on discoveries in genetic research and show great promise. With the completion of the Human Genome Project's mapping of thirty thousand genes, genomic maps will be available to help determine the genetic basis of XP and, we can hope, all skin diseases. Dermatologists will soon be able to determine a patient's response to treatment and chart his or her prognosis with greater efficiency. The twenty-first century will be the "genetic century" for dermatology (and all of medicine), as we discover how mutations bring on skin disease and the multiple mechanisms surrounding their expression.

For melanoma, clinical trials are under way in hopes of replacing traditional chemotherapy, and many specialized oral therapies should soon reach the market, including a new generation of targeted therapies. These drugs attack the underlying molecular mechanism of melanoma, blocking a chain reaction inside the cancer cells that allows them to multiply and attract blood vessels for growth. Use of these new drugs will allow the cancer to be treated as a chronic disease similar to high blood pressure, diabetes, or depression, and advances in melanoma vaccines may help lessen the risk of the deadly cancer.

In the future of dermatology, the causes and effects of human papillomavirus (HPV) and other viruses contributing to skin cancer will be further investigated and tested to help target treatments. A high incidence of HPV is detected in precancers (actinic keratoses, or AKs) and in invasive squamous cell cancers (SCCs). HPV is found in 40 to 50 percent of AKs and in up to 70 percent of SCCs, and HPV incidence is higher in immunocompromised patients. Triple risk factors for developing skin cancers include exposure to UV light, presence of HPV, and dysfunction of the local immune system. In the case of the rare but deadly Merkel cell cancer, another virus, the polyoma virus, has been discovered as its cause. Given this new knowledge and our continuing accumulation of fresh insights into the etiology of skin cancers, we hope in the future to devise a whole array of novel ways to save our skin.

Future scientific discoveries may also demonstrate new antibody-mediated connections for many dermatologic diseases that we have long suspected to have this autoimmune etiology. Through a mixture of good clinical observation, hard work, and luck, dermatological researchers will continue to make connections, and then study whether the phenomena they are observing are only associated or actually causally related to disease formation. We may soon look at the age of dermatological surgery for skin cancers with a healthy nostalgia, when immune therapies and vaccines replace the need for these difficult, time-consuming surgeries.

SKIN AS TECHNOLOGY, TECHNOLOGY AS SKIN

The skin's crucial body-covering role is becoming increasingly recognized, including an enhanced awareness of when the skin can use an outside boost. With most trauma the skin rebounds, but for burn victims who have lost more than 40 percent of their skin surface, a meshwork of donor human skin and grafts are being designed to provide a lasting substitute. New creations include artificial matrices (generally a collagen scaffold) that grow skin from stem cells taken from the foreskin or umbilical cord of newborn infants, and a three-dimensional matrix composed of a combination of human skin cells and biodegradable polymers. The devices act as a foundation for dermal cells to grow and provide a barrier against infection, water loss, and excessive ultraviolet light.

For those with extensive burns, alternatives to skin grafts using sheets of cells cultured from a patient's own skin continue to be developed. In the past, the weeks it took to grow the cells would allow life-threatening infections to interfere with the patient's healing. Today, faster methods that involve speed-culturing skin cells and spraying them onto wound surfaces have been developed. In the future, advances in gene therapy will be combined with speedculture methods to create quick, naturally laminated substitutes for skin.

Electronic skin that mimics the pressure and temperature sensors in our skin is coming but will take time to blossom. As Marc Lappe writes, "The exquisite sensitivity, natural stretchiness, and pliability of human skin are tricky to replicate. Advances in the development of artificial electronic skin for use in robotics have been slow in coming because the complex and economical design of real skin allows it to do its many jobs with only a few millimeters of thickness."

When it is developed, artificial skin will have a multitude of uses. A *New York Times* article suggested it might be useful in testing cosmetics, including perfumes. Other researchers have developed a material that may allow patients to forgo daily injections and pills and receive prescriptions instead through micro-thin implantable films that release medication such as insulin according to changes in temperature and internal demand.

Long-lasting skin fillers are also being studied for their ability to more permanently repair defects and make changes. Already face transplants have been performed and more are coming, a radical procedure intended for patients with severe disfigurement. Although doctors in the past have successfully reattached faces to patients after accidents, transferring facial tissue and blood vessels from a cadaver to a new patient is virgin territory. These transplants bring a lifetime dependence on expensive immunosuppressant drugs to block rejection of the new tissue, but these operations could offer an improved future for those who suffer disfigurement from severe burns, cancer, or gunshot wounds. Of course, the procedure raises major moral, ethical, and psychological issues.

On a more mundane side, hair growth and transplantation will be safer and individual, artificial-appearing hair plugs will be a historical reference. New and more individualized hair growth drugs will become available. Cloning of individual hair cells will allow an unlimited source of replacement hair.

New devices are being developed to detect skin cancer and other skin maladies using image analysis and computer–assisted diagnosis. The devices scan a skin lesion for abnormal appearance and detect any changes since the lesion was last evaluated to allow more well-informed judgments on lesions that might be troublesome and should be removed. This pattern-recognition device is particularly helpful for people with extensive lesions or a family history of melanoma, which increases the possibility of melanoma. Continuous research and refinement will allow improvements in detection and treatment.

Teledermatology, the computer-assisted, long-distance transmission of data about dermatological cases, will bring detection and therapeutic suggestions to areas where hands-on dermatology is limited. Joe Kvedar of Harvard Medical School writes that the Internet "offers the opportunity for delivery of care anytime, anywhere. This care delivery method will enable dermatologists to offer services in a place-independent fashion and may interrupt current referral networks."

Tania J. Phillips, professor of dermatology at the Boston

University School of Medicine, agrees: "I think teledermatology will play an increasing role. Physician extenders will be increasingly used and instruments such as the dermatoscope and other in vivo imaging techniques will be used. Treatments such as the immune response modifier molecules and biologics will be increasingly used for different indications. Hopefully for wound patients there will be new, affordable cell-based therapies available. For education and teaching I think that the Internet and computer-based learning will supplant many of our traditional methods, as they are already doing!"

Although I began my inquiry with the more utilitarian potential of future skin developments, I also realize, given the enormous influence of aesthetics among Homo sapiens, that the future will also include "skin as entertainment."

What if one could change skin colors based on mood? I know of many patients with frustrating blush disorders who wish their state of mind were not so readily visible on their hot red skin. However, others, for example a military person trying to hide from an approaching enemy, might want a change in color. And of course there will be those who suffer a certain ennui from their current, static display of tattoos—and an ever-changing tableau would offer an extensive realm of show and tell.

In the future of skin, the cosmetic world will change dramatically. Tattoos may become even more popular and less expensive. Tattoo removal techniques have already been developed using a single wavelength of visible light to break down the tattoo into harmless and invisible by-products, with little hassle and discomfort. Skin color in the future will become more wide-ranging and stylish, with more natural tans created by products containing melanin, melanin-like pigments, and those that stimulate the body's own melanin production without sun. The reverse will also be true, allowing those who desire to bleach their skin to turn off melanin production at the cellular level. Throw in a mix of colors and you will be able to produce a rainbow of colors for a special event or celebration. And what about the skin as a vehicle for delivery of other drugs besides creams and ointments?

THE FUTURE DERMATOLOGIST

I have often daydreamed about my own profession's future existence. I imagine the not-too-distant future dermatologist's office, devoid of any trace of life as we have known it. Advanced technological devices fill every corner of the space. Patients come and go, often in less time than it takes a laser beam to penetrate a tattoo.

In this dermatological dystopia, pharmaceutical companies are still at work behind the scenes. I can see them now, spending the money we generate for their companies on crucial agendas, such as lobbying the Mars candy company to make M&Ms in titanium dioxide and yellow, to match the forty-milligram tablets of a popular acne medication, thereby smoothing out the transition between eating the candy and swallowing the acne medicine.

Meanwhile, the dermatologists of the future are basking in the light of their computer screens, teledermatology having taken over much of direct patient care. (The new slogans,

"If you were a dermatologist, you could be home by now," or "Why not live like a radiologist? Never leave the comforts of your own home.") The pharmacists and mid-level practitioners will be providing vaccine injections for skin cancers and most dermatological diseases. Of course, the future won't necessarily be so grim. As the dermatopathologist Mike Morgan told me:

> The dermatologist of the late twenty-first century will assume a greater degree of responsibility for diagnosis. Armed with hand-held spectrophotometric and chemical detection devices, the vast majority of cutaneous neoplasms will not only be accurately identified but risk assessed in situ. Characteristic light diffraction spectra will differentially fingerprint the types of cutaneous malignancy and the application of light- or sound-emitting devices will precisely gauge the depth of tumor penetration. Chemical detecting devices programmed to recognize subtle changes in the metabolic by-products of cancerous cells will complement the light-emitting devices. Similarly, these devices will be relied upon to assess the extent of residual disease. Computerized algorithms that reconcile the measured variables of epidermal thickness, vascular density, and depth of inflammatory infiltrate with pre-programmed archetypes will also permit the assessment and identification of many dermatoses. Such advances will undoubtedly change the role of and importance of dermatopathology in the equation of dermatologic care. Dermatopathologists would be relegated to the arbitration of equivocal cases or sought in the assessment of confounding data or following incomplete response to therapy.

From an ethical standpoint, dermatologists in the future will have more access to Internet-based information about new products. This could result in dermatologists more often prescribing treatments supported by objective evidence-based data and research findings, and being less influenced by representatives of the pharmaceutical companies. Having access to this information will help physicians to make their own decisions and not be influenced as much by "drug reps" who wish us to sway prescription-writing choices.

The next generation of dermatologists will be more likely to integrate many disciplines in their work. The future of integrative therapies in dermatology, in particular preventive medicine, botanicals including antioxidants, hypnosis, and behavioral modification will allow new detection and treatment options. Based on research in integrative medicine, new educational and patient teaching options also will be utilized in dermatology.

The future should also be good for dermatological detectives. We may have skin detective agencies utilizing bacteriological forensic techniques, pointing to individuals at the scene of a crime. The characteristic microflora of a criminal suspect could be just as important for identification as a fingerprint or other genetic markers. If an individual's microflora, established shortly after birth from contact with the mother, hospital, and other early presences, remains comparatively constant throughout life, a microbial sampling of room dust, saliva, and other evidence could reveal groups of identifiable organisms that would match the pattern of a particular suspect. Using these sophisticated identification methods, bacteria could be used to give evidence in court. And more obvious skin identification could be important in the future as well: biochips implanted into the skin to transmit personal and medical information will surely be fodder for legal and scientific inquiry.

Should we not put our efforts into having a modicum of health for everyone? Or should we put our energy into the ephemeral yearnings of life for those who can afford all the beauty that the needle and laser can provide us? It is a personal decision perhaps more so than ethical, but I think it is one that each of us must consider.

Hawthorne's *The Birth-Mark* depicts the life of a man named Aylmer who is obsessed with finding perfection in his wife and eliminating her facial birthmark, only to make a horrible mistake.

> Alas! it was too true! The fatal hand had grappled with the mystery of life, and was the bond by which an angelic spirit kept itself in union with a mortal frame. As the last crimson tint of the birthmark—that sole token of human imperfection—faded from her cheek, the parting breath of the now perfect woman passed into the atmosphere, and her soul, lingering a moment near her husband, took its heavenward flight. Then a hoarse, chuckling laugh was heard again! Thus ever does the gross fatality of earth exult in its invariable triumph over the immortal essence that, in this dim sphere of half development, demands the completeness of a higher state. Yet, had Aylmer reached a profounder wisdom, he need not thus have flung away the happiness which would have woven his mortal life of the selfsame texture with the celestial. The momentary circumstance was too strong for him; he failed to look beyond the shadowy scope of time, and, living once for all in eternity, to find the perfect future in the present.

When dermatologists can no longer differentiate a Malpighian from a melanocyte, their differential diagnosis skills atrophied, having faded into distant reservoirs of their cerebral cortices, which slowly dry up under the cosmetic sun, when they all have become advanced aestheticians with limited academic foundation to serve as a springboard for intellectual endeavors, when the knowledge creeps in that their fundamental skills have evaporated like yesterday's cryotherapy spray, when their brains are too scrambled with the selling points of their lotions and creams and snake oils, when they can't tolerate being on the ephemeral see-saw of what is cosmetically fashionable, when the fact that their upcoming laser machine bills are coming due gains purchase in their mind . . . perhaps they will awaken to the fact that their inner Aylmer has taken over. When what we think of now as exotic becomes mediocre, it may be too late to turn back.

Perhaps the cautionary message for the newly trained, freshly minted dermatologist should be "wrinkles are filled, knowledge fades." Dermatologists spend years examining and studying the human body, weaving intricate tapestries of shave-sharp therapeutic acumen and piecing together patterns of signs and symptoms. They are the inheritors of predecessors who devoted years to patient care and wrote down their findings. And they are the beneficiaries of researchers who diligently struggled so that today's physicians can jot down remedies on prescription pads and get free CDs at the medical meetings. But the respect and honor that dermatology and dermatologists should possess is in danger of being crushed under the crunch of cosmetic technology's tires.

For my own part, I have often turned away potential patients who come to me wanting to "fix" their image issues. Red flags often rise high while I talk with patients who request procedures to please someone other than themselves, have ideals of social perfectionism, express a morbid dissatisfaction that signals body dysmorphic syndrome or other deep emotional scars, or hold generally unrealistic expectations and yearn for cosmetic cures for their unhappiness. Botox and chemical peels may improve your self-esteem but they rarely save a floundering marriage.

Of course, many patients come to dermatologists for help with disturbing medical problems. Often they have significant maladies—aggressive skin cancers, horrible acne, lupus, or miserable psoriasis—and medically oriented dermatologists, myself included, try to care for these souls in need of repair.

I am not a cosmetic Luddite. I've botoxed unwanted facial lines, frozen a number of irritated blemishes, and made spider veins (superficial blood vessels) disappear. But everything in moderation. I believe that as a dermatologist, my medical knowledge separates me and my staff from this country's countless Fountain of Youth Cosmetic clinics, which may not have adequately trained personnel. When it comes to the tough medical cases, leave it to those with the extensive training required to do the job.

I can certainly see the temptation: cosmetic payment might be as primitive an exchange as you can have—here's what I do for you and here's how you pay me. It's easy to contrast that simplicity with the hassle of dealing with the HMOs, national health insurance plans, and lawyers who often interpret our good intentions in medical care with the subtlety of someone going after a mosquito with an axe. But the great medical teachers were aware that some problems, like the insecurities that cause so many to seek cosmetic procedures, can't be solved. In his last aphorism (Section VII, number 87), Hippocrates writes, "Those diseases which medicines do not cure, iron [the knife?] cures; those which iron cannot cure, fire cures; and those which fire cannot cure, are to be reckoned wholly incurable." Rather than risk a negative outcome for an otherwise benign cosmetic condition, it is at times better to leave it alone, as Hippocrates implies in perhaps his most famous quote, "Life is short, and Art long; the crisis fleeting; experience perilous, and decision difficult." Given all the major obstacles during his time, he certainly perceived that attempts at cosmetic perfection were frivolous. As for Sir William Osler, the great physician, diagnostician, and one of the founders of the Johns Hopkins hospital, he has got to be palpating himself in his grave, wondering what has happened to diagnostic and clinical skills. Dermatologists and other physicians have moved away from diagnostic acumen based on knowledge gained from doing all types of hands-on and auditory detection—including listening to patients' stories—in favor of ordering elaborate and exorbitant tests for too many initial complaints, as a sort of diagnostic short-cut. "Listen to your patient, he is telling you the diagnosis," Sir William Osler often wrote, emphasizing the importance of taking a good history.

It's not only that dermatologists can't always fix cosmetic imperfections—it's that these imperfections are part of the variety that makes human life interesting. In *Nature's Chaos*, James Gleick writes, "The essence of the earth's beauty lies in disorder, a peculiarly patterned disorder, from the fierce tumult of rushing water to the filigrees of unbridled vegetation." Gleick, in his fervor for fractals and chaos theory, states that the "motivating contention of the new science of chaos theory is that such seeming irregularities can be contemplated, sorted, measured, and understood. Traditionally scientists looked for a more conventional order in nature and treated the erratic as a side issue, an unpredictable and therefore unimportant kind of marginalia. Now scientists are more willing to look directly at the irregularity." In similar fashion, when we try to tame any perceived human disorder (including birthmarks) we must have some inherent sense of orderliness in our vision and perception of nature. However, I submit that the essence of human beauty lies in its unpredictability, which draws our divine attention, an irreverent, captivating punch line almost imperceptibly delivered.

"We're so busy finding out if we *could* that we never stopped to find out if we *should*," the character Dr. Ian Malcom states in the movie *Jurassic Park*. I find this statement a great summary of how dermatology has plowed ahead, trying to discover remedies for life's built-in deformities, rarely pausing to question the price in both economic and life-quality terms.

What all striving misses is that without sickness we cannot appreciate our golden health. I doubt there is a reader who has not rejoiced at her or his own revival after a bout of sickness or trauma. For myself, I cherish the little imperfections of life, the physical and mental foibles that make us humble, make us laugh, and make us the most fully human and wishing to tell our stories.

As far as I know, we are the only species to have dermatologists, nail salons, beauty parlors, and myriad other sources to rid our body of real or perceived ailments. I am forever humbled, for along with my fellow soldiers who fight these everlasting skin diseases, I know we can never win the battle. However, when I think about the thousands of patients I treat with skin problems every year, I'm glad I can provide a little help along the way and I'm looking forward to the future and what more I and my colleagues can offer.

LIVING WITHIN OUR SKINS

Skin is truly a wonder: an information source and sensory receptor, a temperature and blood pressure regulator, a protective barrier, an immunologic source of hormones for protective cell differentiation, and a compound synthesizer. To maintain this amazing covering you have been given, you have choices. You should now be aware of the dangers of tanning, too much sun exposure, and other harmful behaviors.

Skin affects the way we see each other, but living below that surface is a world that is seldom seen or discussed. We host an enormous number of hidden creatures who in general are quite harmless if a proper balance is maintained. In this way our skin mimics our surroundings, and we share certain skin maladies with plants and other animals.

Skin color is one of the central determiners of our lives from a social and cultural perspective. Each day I witness the power of the melanocyte and how much it can determine in our life, especially when a deficit occurs as with vitiligo or albinism.

As the stories that focus on clinical tales reflect, many patients endure skin conditions with a patience that ranges from stoic to heroic. While they are enduring clinical investigations, treatments, and waiting on a name for their malady, they carry on with the routines of daily living whenever possible.

Many dermatological conditions are associated with a psychiatric disorder, and these body-mind disorders often bring us to battle against our own skin. Patients may suffer with disabling skin hypochondriasis. I have explored how aggravating psychosocial factors can mix with an underlying personal or family history of psychiatric and skin illness to ignite a fire that is highly resistant to being extinguished.

I have been a witness to so many people suffering from skin diseases, people sitting inside a covering of distress, and have often hypothesized about the etiology of their problems and observed the patterns of their clinical presentations. Many eruptions are exacerbated by precipitant stress and life changes. In some cases the pathology is isolated and random. In others it is a violent rainstorm, as with inflammation that affects nearly the entire skin surface in erythema and scaling, or the peeling and blistering of bullous disease. All of these appear in an otherwise quiet and clear exam room, with various observers coming and going.

Beyond the exam room, a million other stories play out. The child kept at home because of a fungal hair disease or a proliferating viral disease such as molluscum contagiosum. The high school student who chooses not to go to school or engage in social activities because of the embarrassment of widespread psoriasis or severe acne. The mother of an infant with atopic dermatitis who has to leave her job because she cannot function due to lack of sleep, as her child scratches, cries, and keeps her awake for most of each night. The older man who cannot bear to have yet another cancer, and fears going to find out about the itchy and bleeding growth on his scalp.

Because skin is our boundary with the outer world, those with severe skin disease can suffer socially devastating consequences. The vitiligo patient may be excluded from school and other activities due to ignorance and racial prejudice. The pain of severe itching and scaling from psoriasis often is compounded by the unfounded observation of others that the disease is contagious and contact should be avoided. The number and variety of skin afflictions can only be matched by the inventiveness of those trying to cope with their disconcerting roles as involuntary hosts.

People with skin diseases may find themselves not only perceived as culturally abnormal and stigmatized but offered limited opportunities and socially marginalized. Given our culture's Hollywood glamour mentality, the skinny, unblemished models on every beauty-oriented magazine cover, and the unrealistic standards we often use to feed our self-esteem, those with skin diseases can often barely cope.

Certain skin diseases clearly can exact a devastating toll. Healing the psychological and physical pain of my patients are the chief goals of my daily work. Each day I explore and learn more about how the skin and our perception of it influence our social, cultural, spiritual, and physical being. I hope that that inquiry, and this book, has helped you to learn and grow in your own skin.

GLOSSARY

ALOPECIA: Hair loss.

ANNULAR: Ring shaped.

ASTEATOTIC: Excessively dry.

BULLAE: Large blisters.

CARBUNCLE: The merging of several adjacent furuncles.

COMEDONES (singular, comedo): White, gray, or black plugs in the pilosebaceous openings (each opening has a hair follicle, sebaceous gland, and an arrector pili muscle), consisting of dried sebum, cellular debris, and bacteria; they include blackheads and whiteheads.

CRUSTS: Dried fluid. Black crusts are usually from blood. Yellow crusts represent dried serum, as from bullous lesions. Brownish or honey-colored crusts are secondarily infected with bacteria, as in impetigo.

CYSTS: Noninflammatory collections of fluid or semisolid substances surrounded by a well-defined wall.

DEPIGMENTED: Lacking pigment entirely (e.g., vitiligo).

ECCHYMOSES: Large areas of bleeding into the skin.

ECZEMATIZATION: A combination of weeping, oozing, vesiculation, erythema, crusting, and lichenification.

ERYTHEMA: Reddening of the skin; it is generally associated with an increase in blood flow to an area.

EXCORIATIONS: Scratch marks.

FOLLICULITIS: Inflammation of hair follicles.

FURUNCLES: Large and deep infections of hair follicles.

GEOGRAPHIC TONGUE: A benign inflammatory condition of the mucous membrane of the tongue, in which the absence of papillae makes the tongue look like a map, with smooth red areas resembling the islands of an archipelago.

GUTTATE ERUPTIONS: Lesions that are small and in the shape of drops.

HIRSUTISM: In women and children, an adult male pattern of excessive, androgenic body hair on the face, back, or chest.

HYPERPIGMENTED: Having an increase in the normal melanin pigment.

HYPERTRICHOSIS: Abnormal amount of hair growth over the body, in localized or general distribution.

HYPOPIGMENTED: Having a decrease in pigment.

INTERTRIGINOUS SKIN: An area of skin where adjacent skin surfaces rub against each other and moisture is trapped, such as the underarm and groin.

IRIS LESIONS: Lesions that have a central point with a ring around it.

LICHENIFICATION: Exaggerated skin markings.

LINEAR: Lesions in a line.

KERATOSIS: A circumscribed increase of the horny layer made up of keratin, the major protein in the epidermis.

MACERATION: Continuous wetting of the skin, which produces thickening and whitening of the skin.

MACULE: A lesion in which the only abnormality is a change in color.

MACULOPAPULAR ERUPTIONS: Eruptions that have some areas of only erythema and other areas of erythematous papules.

NODULES: A raised or indurated lesion, slightly larger than a papule.

PAPULES: A raised or indurated lesion generally less than one centimeter in diameter.

PAPULOSQUAMOUS: Raised lesions with scaling on their surfaces.

PAPULOVESICULAR ERUPTION: Eruptions that have both papules and vesicles.

PATCHES: Areas of color change larger than about one centimeter in diameter.

PETECHIAE: Small hemorrhages from superficial blood vessels, which therefore do not blanch.

PLAQUE: An elevated, plateau-like lesion, which develops from the coalescence of smaller lesions, such as wheals or papules.

PUSTULE: A cavity filled with pus, which is made up of a mixture of fluid, cellular debris, and microorganisms.

SCALE: A dried-out bit of excess horny material. These may be thick, fine, or forming a circle around the periphery of the lesion; also called a squame.

SERPIGINOUS: Lesions that wind in a snake-like pattern.

TARGET LESIONS: Lesions that have a central point with a ring around it and are usually seen in erythema multiforme.

TELANGIECTASES (singular, telangiectasia): Permanently dilated, small, superficial blood vessels.

TUMOR: A very large nodule.

VEGETATION: A large, moist, cauliflower-like growth.

VERRUCOUS: Wartlike skin surface changes.

VESICLES: Small blisters.

WHEAL: The medical term for a hive, also called urticaria.

XEROTIC: Excessively dry.

BIBLIOGRAPHY

Aghel, Azadeh, and Amor Khachemoune. "What Is This Brown Papule?" *The Dermatologist* 19, no. 1 (January 2011): 47–50.

Allen, Karen. "Living in Fear: Tanzania's Albinos." BBC.com, 21 July 2008.

Andrews, Michael. *The Life That Lives on Man*. Taplinger, 1977

Balin, Arthur K., and Loretta Pratt Balin. *The Life of the Skin: What It Hides, What It Reveals, and How It Communicates*. New York: Bantam Books, 1997.

Batai, I., S. Marton, R. Batai, and M. Kerenyi. "Single Dose Intravenous Cephazolin Prophylaxis Rapidly Reduces Normal Skin Flora." *European Journal of Anaesthesiology* 21 (June 2004): 137.

Bieber, L. "Staphylococcus Lugdunensis in Several Niches of the Normal Skin Flora." *Clinical Microbiology and Infection* 16, no. 4 (April 2010): 385–88.

Boniol, M. "Cutaneous Melanoma Attributable to Sunbed Use: Systematic Review and Meta-analysis." *British Medical Journal* 345 (24 July 2012).

Carvajal, D. "A New Science, at First Blush." *New York Times*, 20 November 2007.

Chang, A., J. Wong, J. Endo, and R. Norman. "Geriatric Dermatology: Part II, Risk Factors and Cutaneous Signs of Elder Mistreatment for the Dermatologist." *Journal of the American Academy of Dermatology* 68, no. 4 (April 2013).

————. "Geriatric Dermatology Review: Major Changes in Skin Function in Older Patients and Their Contribution to Common Clinical Challenges." *Journal of the American Medical Directors Association* 14, no. 10 (October 2013): 724–30.

Chen, T. "Clinical Values of Mild Cleansers That Are Effective in Reducing Skin Flora." *Journal of the American Academy of Dermatology* 54, no. 3 (March 2006): 108.

CITY100 (Controlling Indoor Tanning in Youth). Website. 2008. http://indoortanningreportcard.com/ourfindings.html.

Cohen, Russell. "Tanning Trouble." *Scholastic Choices* 18, no. 4 (31 January 2003): 23.

Cokkinides, V., M. Weinstock, D. Lazovich, E. Ward, and M. Thun. "Indoor Tanning Use among Adolescents in the US, 1998–2004." *Cancer* 115, no. 1 (2009): 190–98.

"Cosmetic Tattooing." *Illawarra Mercury*, 3 November 2011.

Darwin, Charles. *The Descent of Man, and Selection in Relation to Sex.* London: John Murray, 1871.

de Almeida e Borges, L. F., B. L. Silva, and P. P. Gontijo Filho. "Hand Washing: Changes in the Skin Flora." *American Journal of Infection Control* 35, no. 6 (January 2007): 417–20.

Dellavalle, R. P., E. R. Parker, N. Ceronsky, E. J. Hester, B. Hemme, D. L. Burkhardt, A. K. Chen, and L. M. Schilling. "Youth Access Laws: In the Dark at the Tanning Parlor?" *Archives of Dermatolology* 139, no. 4 (2003): 443–48.

Dohi, S. "Tattooing and Syphilis." *Archives of Dermatology* 146, no. 3 (1960): 233.

Dron, P. "Allergy to Piercing and Tattooing." *Revue Française d'Allergologie et d'Immunologie Clinique* 47, no. 6 (October 2007): 398–401.

Endo, J., J. Wong, R. Norman, and A. Chang. "Geriatric Dermatology: Part I. Geriatric Pharmacology for the Dermatologist." *Journal of the American Academy of Dermatology* 68, no. 4 (April 2013).

Fisher, D. E., and W. D. James. "Indoor Tanning: Science, Behavior, and Policy." *New England Journal of Medicine* 363 (2010): 901–3.

Fisher, R. E.. "The Primate Appendix: A Reassessment." *The Anatomical Record (New Anatomist)* 261 (2000): 228–36.

Fried, R. G. "Evaluation and Treatment of 'Psychogenic' Pruritus and Self-Excoriation." *Journal of the American Academy of Dermatology* 30, no. 6 (1994): 993–99.

Fried, R., and S. Fried. "Picking Apart the Picker: A Clinician's Guide for Management of the Patient Presenting with Excoriations. *Cutis* 71, no. 4 (April 2003): 291–98.

Fundukian, Laurie J. "Tanning." In *Gale Encyclopedia of Medicine*, 4247–50. Detroit: Gale, 2011.

Gale, Jason. "Tanning Beds Particularly Dangers for Teens Says New Study." *Bloomberg News* (Melbourne), 4 December 2012.

Gallagher, James. "Sunless Tanning Booths Draw New Customers." *Augusta Chronicle*, 18 January 2004.

Gleick, J., and E. Porter. *Nature's Chaos*. Boston: Little, Brown and Company, 2001.

Gloster, H. M., and K. Neal. "Skin Cancer in Skin of Color." *Journal of the American Academy of Dermatology* 55, no. 5 (2006): 741–60.

Goldsmith, Connie. "What's Your Tanning IQ?" *Current Health 2* 25, no. 8 (April/May 1999): 22.

Gould, G., and W. Pyle. *Anomalies and Curiosities of Medicine*. Philadelphia: W. B. Saunders, 1896.

Gupta, Deepak. "Cost-effective Tattooing." *Journal of Glaucoma* 19, no. 8 (January 2010): 566–67.

Gupta, M. A., and A. K. Gupta. "Fluoxetine Is an Effective Treatment for Neurotic Excoriations: Case Report." *Cutis* 51, no. 5 (1993): 386–87.

Gyorimolnar, I., S. Czuczor, R. Batai, M. Kereni, and I. Batai. "The Impact of Preoperative Total Body Showering with Povidone-Iodine on Skin Flora." *European Journal of Anaesthesiology* 22, no. 35 (June 2005): 46.

Hadaway, Lynn C. "Skin Flora and Infection." *Journal of Infusion Nursing* 26, no. 1 (January 2003): 44–48.

———. "Skin Flora: Unwanted Dead or Alive." *Nursing* 35, no. 7 (July 2005): 20.

Hakim, E. "Minocycline-Induced Pigmentation: Incidence, Prevention and Management." *Drug Safety* 18, no. 6 (June 1998): 431–40.

Heckman, Carolyn J. "Prevalence and Correlates of Indoor Tanning among U.S. Adults." *Journal of the American Academy of Dermatology* 58, no. 5 (May 2008): 769–80.

Hoerster, K., et al. "Density of Indoor Tanning Facilities in 116 Large U.S. Cities." *American Journal of Preventive Medicine* 36, no. 3 (March 2009): 243–46,.

Holman, D. M., K. A. Fox, J. D. Glenn, G. P. Guy, M. Watson, K. Baker, V. Cokkinides, M. Gottlieb, D. Lazovich, F. M. Perna, B. P. Sampson, A. Siedenberg, C. Sinclair, and A. C. Geller. "Strategies to Reduce Indoor Tanning: Current Research Gaps and Future Opportunities for Prevention." *American Journal of Preventive Medicine* 44, no. 6 (June 2013): 672–81.

Hornung, R. L., K. H. Magee, W. J. Lee, L. A. Hansen, and Y. C. Hsieh. "Tanning Facility Use: Are We Exceeding the Food and Drug Administration Limits?" *Journal of the American Academy of Dermatology* 49, no 4 (October 2003): 655–61.

Jablonski, Nina. *Skin: A Natural History.* Berkeley: University of California Press, 2006.

James, M. D., C. J. Cockerell, L. M. Dzubow, et al., eds. *Advances in Dermatology*, vol. 15. St. Louis: Mosby Publishing, 2000.

Karagas, M., V. A. Stannard, L. A. Mott, M. J. Slattery, S. K. Spencer, and M. A. Weinstock. "Use of Tanning Devices and Risk of Basal Cell and Squamous Cell Skin Cancers." *Journal of the National Cancer Institute* 94, no. 3 (6 February 2002): 224–26.

Kethu, Sripathi R. "Endoscopic Tattooing." *Gastrointestinal Endoscopy* 72, no. 4 (October 2010): 681–85.

Koblenzer, C. S. "Cutaneous Manifestations of Psychiatric Disease That Commonly Present to the Dermatologist: Diagnosis and

Treatment." *International Journal of Psychiatry in Medicine* 22 (1992): 47–63.

Koerting, Katrina. "Study: Using Tanning Beds Is a Cancer Risk." *Connecticut Post* (Bridgeport) 4 August 2009.

Koo, J. "Psychodermatology: A Practical Manual for Clinicians." *Current Problems in Dermatology* 7 (1995): 204–32.

Kvedar, J., personal conversation and e-mail.

Kwon, H. T., J. A. Mayer, K. K. Walker, H. Yu, E. C. Lewis, and G. E. Belch. "Promotion of Frequent Tanning Sessions by Indoor Tanning Facilities: Two Studies." *Journal of the American Academy of Dermatology* 46, no. 5 (2002): 700–70.

Lappe, Marc. *The Body's Edge: Our Cultural Obsession with Skin.* New York: Henry Holt, 1996.

Lapresta, A., C. Pérez, and D. García-Almagro. "Facial Lesions after Tattooing." *Actas Dermo-Sifiliográficas* 101, no. 10 (January 2010): 889–90.

Larson, E. L., A. B. Cronquist, L. Lai, C. T. Lyle, and P. Della Latta. "Differences in Skin Flora between Inpatients and Chronically Ill Outpatients." *Heart and Lung* 29, no. 4 (January 2000): 298–305.

Lazovich, D., R. I. Vogel, M. Berwick, M. A. Weinstock, K. E. Anderson, and E. M. Warshaw. "Indoor Tanning and Risk of Melanoma: A Case-Control Study in a Highly Exposed Population." *Cancer Epidemiology, Biomarkers and Prevention* 19, no. 6 (June 2010): 1557–68.

McElroy, S. L., J. I. Hudson, K. A. Phillips, P. E. Keck, and H. G. Pope. "Clinical and Theoretical Implications of a Possible Link between Obsessive-Compulsive and Impulse Control Disorders." *Depression* 1, no. 3 (1993): 121–32.

Mempel, M. "News on Skin Infection Research for GPs: The Role of Physiological Skin Flora in Protection against Pathogens." *Journal der Deutschen Dermatologischen Gesellschaft* 9 (April 2011): 20.

Miller, S. A., S. L. Hamilton, U. G. Wester, and W. H. Cyr. "An Analysis of UVA Emissions from Sunlamps and the Potential Importance for Melanoma." *Photochemical and Photobiological Sciences* 68, no. 1 (1998): 63–70.

Montagu, Ashley. *Touching: The Human Significance of the Skin.* 3rd edition. New York: Harper and Row, 1986.

Morgan, M., personal conversation and e-mail.

Nagoshi, Julie, and Craig Nagoshi. "Plastic Surgery, Tattooing, and Piercing." In *Battleground: Women, Gender, and Sexuality,* edited by Stephanie Brzuzy and Amy Lind, 389–95. Westport, CT: Greenwood, 2008.

Norman, R. *Geriatric Dermatology.* New York: Parthenon, 2001.

———. *Diagnosis of Aging Skin Diseases.* London: Springer-Verlag, 2008.

———. *100 Questions and Answers on Aging Skin.* Boston: Jones and Bartlett, 2009.

———. *Preventive Dermatology.* London: Springer-Verlag, 2010.

———. *Endo J Clinical Cases in Geriatric Dermatology.* London: Springer-Verlag, 2012.

Norman, R., and L. Reuscher. *100 Questions and Answers on Chronic Illness.* Boston: Jones and Bartlett, 2010.

Nuland, Sherwin. *The Soul of Medicine: Tales from the Bedside.* New York: Kaplan, 2010.

Nurse, Earl. "We Are Killed, We Are Hunted: Albino Activist Fights Witchcraft Murders." CNN.com, 17 May 2013.

Osler, William. *The Principles and Practice of Medicine: Designed for the Use of Practitioners and Students of Medicine.* Edinburgh: Young J. Putland, 1892.

Parsad, Davinder, Sunil Dogra, and Amrinder Jit Kanwar. "Quality of Life in Patients with Vitiligo." *Health and Quality of Life Outcomes* 1 (2003): 58.

Pendergast, Sara, and Tom Pendergast. "Tattooing." In *Fashion, Costume, and Culture: Clothing, Headwear, Body Decorations, and Footwear through the Ages,* 244–46. Detroit: Thomson-Gale, 2004.

Phillips, K. A., and S. L. Taub. "Skin Picking as a Symptom of Body Dysmorphic Disorder." *Psychopharmacology Bulletin* 31, no. 2 (1995): 279–88.

Phillips, T., personal conversation and e-mail.

Pomeranz, J. "Does the Appendix Serve a Purpose in Any Animal?" *Scientific American,* 24 August 2001.

Raj, S. Dorai. "Only Skin Deep." *The Hindu* 28, no. 3 (2011).

Reynolds, Diane. "Literature Review of Theory-Based Empirical Studies Examining Adolescent Tanning Practices." *Dermatology Nursing* 19, no. 5 (October 2007): 440.

Sato, S. "Human Skin Flora as a Potential Source of Epidural Abscess." *Anesthesiology* 85, no. 6 (December 1996): 1276–82.

Sawyer, S. *Body Piercing and Tattooing: The Hidden Dangers of Body Art.* New York: Rosen, 2008.

Sebastian, G. "Conduction Anesthesia for Tattooing." *Der Anaesthesist* 55, no. 7 (July 2006): 808.

Selzer, Richard. *Mortal Lessons: Notes on the Art of Surgery.* New York: Simon & Schuster, 1976.

Setlur, Jennifer. "Cosmetic and Reconstructive Medical Tattooing." *Current Opinion in Otolaryngology and Head and Neck Surgery* 15, no. 4 (August 2007): 253–57.

Stein, D. J., C. S. Hutt, and J. L. Spitz. "Compulsive Picking and Obsessive-Compulsive Disorder." *Psychosomatics* 34, no. 2 (1993): 177–81.

Stöppler, Melissa, ed. "Vitiligo Treatment, Symptoms, Causes, Diagnosis and Signs." MedicineNet.com. Last accessed 5 October 2013.

Swerdlow, A. J., and M. A. Weinstock. "Do Tanning Lamps Cause Melanoma? An Epidemiologic Assessment." *Journal of the American Academy of Dermatology* 38, no. 1 (1998): 89–98.

Tabin, Clifford. "Why We Have (Only) Five Fingers per Hand: Hox Genes and the Evolution of Paired Limbs." *Development* 116, no. 2 (1992): 289–96.

"Tanning Salons in the US." *Tanning Salons Market Research Report,* 7 August 2009.

Thamlikitkul, Visanu. "Skin Flora of Patients in Thailand." *American Journal of Infection Control* 31, no. 2 (April 2003): 80–84.

Towfigh, S., W. G. Cheadle, S. F. Lowry, M. A. Malangoni, and S. E. Wilson. "Significant Reduction in Incidence of Wound

Contamination by Skin Flora through Use of Microbial Sealant." *Archives of Surgery* 143, no. 9 (September 2008): 885–91.

U.S. Department of Health and Human Services, Public Health Service, National Toxicology Program. "Exposure to Sunlamps or Sunbeds." *Twelth Report on Carcinogens* (10 June 2011): 429.

van der Velden, Eddy M. "Cosmetic and Reconstructive Medical Tattooing." *Current Opinion in Otolaryngology and Head and Neck Surgery* 13, no. 6 (December 2005): 349–53.

Vassileva, Snejina. "Medical Applications of Tattooing." *Clinics in Dermatology* 25, no. 4 (July 2007): 367–74.

Webb, Sam. "Tragic Plight of the 'Tribe of Ghosts.'" *Daily Mail,* 7 August 2013.

Whitmore, S. E., W. L. Morison, C. S. Potten, and C. Chadwick. "Tanning Salon Exposure and Molecular Alterations." *Journal of the American Academy of Dermatology* 44, no. 5 (2001): 775–80.

Wiedersheim, R. *The Structure of Man: An Index to His Past History.* 2nd edition. Translated by H. and M. Bernard. London: Macmillan and Co. (1893) 1895.

Wilson, Codi. "Tanning-Booth Blues: Cancer Society Advises against Base Tan Strategy." *Calgary Herald,* 7 April 2011.

Wood, Sarah. "Sunless Tanning." *Ocala Star-Banner* (Marion County, FL) 14 June 2005.

Yoo, Jane Y., Gary W. Mendese, and Daniel S. Loo. "Black Dot Tinea Capitis in an Immunosuppressed Man." *Journal of Clinical Aesthetic and Dermatology* 5, no. 5 (May 2013): 49–50.

Yuan, S., et al. "Secondary Syphilis Presenting in a Red Tattoo: Correspondence." *European Journal of Dermatology* 20, no. 4 (July–August 2010): 544–45

Zimmerman, Jessica. "Ernest Hemingway and Cats." *History by Zim: Beyond the Textbooks* (blog). Last modified 21 August 2013.

Text:	10.75/15 Janson MT Pro
Display:	Ultramagnetic
Design and composition:	Lia Tjandra
Printer and binder:	Maple Press